T0211876

Springer Texts in Business and Economics

More information about this series at http://www.springer.com/series/10099

Kam Yu

Mathematical Economics

Prelude to the Neoclassical Model

 Springer

Kam Yu
Lakehead University
Thunder Bay
ON, Canada

ISSN 2192-4333   ISSN 2192-4341   (electronic)
Springer Texts in Business and Economics
ISBN 978-3-030-27291-3   ISBN 978-3-030-27289-0   (eBook)
https://doi.org/10.1007/978-3-030-27289-0

This Springer imprint is published by the registered company Springer Nature Switzerland AG.
The registered company address is: Gewerbestrasse 11, 6330 Cham, Switzerland

Preface

Every year a first-year student, out of curiosity or because of some program requirements, takes principles of economics. She is surprised to learn that a subject in social science can be studied in such a logical and scientific manner. After getting a good grade, she decides to major in Economics. The advisor in the department told the student that quantitative analysis is very important in the study of Economics. A degree in Economics, however, is a liberal arts degree, which provides a well-rounded general education and a good preparation for professional schools. Many Economics majors, the advisor said, achieve high scores in LSAT, an entrance examination for the faculty of law.

The student diligently selects courses in calculus, linear algebra, probability and statistics and econometrics and does well in those courses. After graduation, she decides to pursue graduate study in Economics. To the student's surprise, the verbal and graphical analysis she was trained, in order to solve economic problems, somehow all but disappear. Instead, most of the graduate textbooks contain materials that more or less resemble applied mathematics. It seems that there is a wide gap between the general liberal arts degree and graduate studies in Economics.

The objective of this book is to fill such a gap. It is written for a one-semester course as an introduction to Mathematical Economics for first-year graduate and senior undergraduate students. The materials assume the students have backgrounds in college calculus and linear algebra. A course in mathematical economics at the undergraduate level is helpful but not necessary. Chapters 1 through 5 aim to build up the students' skills in formal proof, axiomatic treatment of linear algebra and elementary vector differentiation. Chapters 6 and 7 present the basic tools needed for microeconomic analysis. I have covered much of the materials up to Chap. 7 to such a class, for many years. Due to the time constraint of a short semester, many proofs of the theorems are omitted. For example, although the inverse function theorem and the implicit function theorem are important tools in economic analysis, experience has shown that the marginal benefits of going through the proofs in class are not worth the distraction. Therefore, throughout the text I have included notes that point to references if the students want to explore a bit more about particular topics.

For these reasons, the book is not meant to be a manual or a reference in Mathematical Economics. My focus is to provide a quick training to graduate students in one semester, to acquire a somewhat elusive quality called mathematical maturity. Greg Mankiw advises his students to take as many courses in mathematics "until it hurts". Different students, of course, have different levels of tolerance. There were occasional moments of delight, when a student exclaimed that the proof of a theorem was simply beautiful. Nevertheless, my apology if this book really hurts.

There are two bonus chapters. Most graduate level econometrics courses assume the students to have some training in mathematical probability and statistics. Chapter 8 provides a quick introduction or review to probability theory. It is, however, by no means a substitute for formal training. Chapter 9 is an introduction to dynamic modelling, which I used as a supplementary reading for a course in macroeconomic theory. Both chapters sometimes refer to concepts covered in the core chapters.

Part of the book was written when I was visiting Vancouver, British Columbia. One summer day I joined a tour with my friend to Whistler. While the tour guide was introducing the Coast Mountains and other points of interest, my mind at one moment drifted back to the time when I was new to the province. I saw afresh how beautiful the country was. In writing this book, there were moments that I thought about the materials from a student's perspective and rediscovered the beauty of mathematics. I hope you enjoy the journey.

Last but not least, I would like to thank my friends Eric, Erwin and Renee, who continue to support me in good and bad times.

Thunder Bay, ON, Canada   Kam Yu

Contents

List of Figures

List of Tables

Logic and Proof

1

1.1 Logical Statements

Most languages follow some grammatical rules in order to convey ideas clearly. Mathematics written in English of course has to follow these rules. Moreover, the ideas have to obey logical rules to be meaningful. In this chapter we review the commonly used logical rules in mathematics. We start with the idea of a **statement**, which is a sentence that a **truth value** can be assigned. The truth value is either true (T) or false (F), but not both.

Example 1.1

(a) Aurorae are formed when solar particles interact with the Earth's magnetic field.
(b) The cardinality of the set of real numbers is bigger than that of the rational numbers.
(c) All prime numbers are odd numbers.
(d) Every child should be entitled to a good education.
(e) $x^2 + 1 = 0$.

The first three sentences are statements. Statement (a) is true based on inductive reasoning. That is, it can be verified by repeated empirical testings and observations, which is the foundation of scientific inquiries. Statement (b) is true because of deductive reasoning. Mathematical knowledge is derived from agreements among mathematicians on assumptions called axioms. Theorems and propositions are derived from the axioms by logical rules. Therefore pure mathematics in principle is not science since it does not rely on empirical observations. Statement (c) is false since 2 is prime but even. The sentence in (d) is a moral declaration. It belongs to the discussion in ethics and cannot be in principle assigned the value of true or false. The last sentence is not a statement because there are some numbers which satisfy the equation, while others do not. We shall later impose more structure on

© Springer Nature Switzerland AG 2019
K. Yu, *Mathematical Economics*, Springer Texts in Business and Economics,
https://doi.org/10.1007/978-3-030-27289-0_1

the sentence to make it a statement. Notice that we do not have to know the truth value before deciding if a sentence is a statement. All we need is that a truth value can in principle be determined.

1.2   Logical Connectives

Logical connectives are operations on statements that change their truth values or combine them to form new statements. If p is a statement, the **negation** of p, written as $\sim p$, has the opposite truth value of p. That is, if p is true, then $\sim p$ is false. If p is false, then $\sim p$ is true. The operation can be illustrated by a **truth table**:

p	$\sim p$
T	F
F	T

For example, if p is "Today is Sunday", then $\sim p$ means "Today is not Sunday". You should convince yourself that $\sim (\sim p)$ has the same truth value as p.

If p and q are two statements, the **conjunction** of p and q, also called "p and q" and denoted by $p \wedge q$, is a statement that is true when both p and q are true. Otherwise it is false. The following truth table lists the four cases:

p	q	$p \wedge q$
T	T	T
T	F	F
F	T	F
F	F	F

The **disjunction** of p and q, also called "p or q" and denoted by $p \vee q$, is a statement that is true when at least one of p and q is true. The truth table is

p	q	$p \vee q$
T	T	T
T	F	T
F	T	T
F	F	F

Our definition of disjunction is sometimes called "inclusive or". In other applications such as computer science, an "exclusive or" means that the disjunction is true if only one of p and q is true, but not both.

Notice that the order of p and q is not important in conjunction and disjunction. That is, "p and q" is the same as "q and p". This is not true for an **implication**, which is written as "If p then q", and denoted by $p \Rightarrow q$. The statement p is called the **hypothesis** and q the **conclusion**. An implication is also called a conditional statement. The rule is defined in the following truth table:

p	q	$p \Rightarrow q$
T	T	T
T	F	F
F	T	T
F	F	T

That is, an implication is false when the hypothesis is true but the conclusion is false. When the hypothesis is false, the implication is true whether the conclusion is true or false.

Example 1.2   Consider the implication "If I am the queen of England, then I would make you a knight". Since the hypothesis is false (I am not the queen), whether I keep my promise or not is irrelevant. Therefore I do not make a false statement.

Implication is one of the most common forms of logical statements. There are many different ways to express the same idea. The following statements all mean "If p then q".

- p implies q.
- q whenever p.
- q if p.
- p only if q.
- q is the necessary condition of p.
- p is a sufficient condition for q.

Given an implication $p \Rightarrow q$, it is sometimes useful to define the following induced conditional statements:

Converse: $q \Rightarrow p$.
Inverse: $(\sim p) \Rightarrow (\sim q)$.
Contrapositive: $(\sim q) \Rightarrow (\sim p)$.

Two statements p and q are **equivalent statements** if p implies q and q implies p. That is, $(p \Rightarrow q) \wedge (q \Rightarrow p)$, which has a shorthand notation $p \Leftrightarrow q$. As the following truth table shows, this happens when p and q have the same truth value.

p	q	$p \Rightarrow q$	$q \Rightarrow p$	$p \Leftrightarrow q$
T	T	T	T	T
T	F	F	T	F
F	T	T	F	F
F	F	T	T	T

In other words, p is equivalent to q when p implies q and the converse is also true. Often we write equivalent statements as "p if and only if q". All definitions are equivalent statements. However, the following example illustrates that we use "if" in definitions when we actually means "if and only if".

Example 1.3 An integer n is odd if it can be expressed as $n = 2m + 1$ where m is an integer. Here the statements p and q are

- p: n is an odd integer.
- q: $n = 2m + 1$ where m is an integer.

Intuitively, equivalent statements have the same meaning and therefore can be used as alternative definitions. Often we call finding equivalent statements **characterization** of the original statement.

1.3   Quantified Statements

In Sect. 1.1 we claim that $x^2 + 1 = 0$ is not a statement. We can turn it into a statement by adding a **quantifier**. For instance, "There exists a real number x such that $x^2 + 1 = 0$" is an example of existential statement. In logical symbols, it reads

$$\exists\, x \in \mathbb{R} \ni x^2 + 1 = 0.$$

The symbol $\exists$ above means "there exists" and is formally called an **existential quantifier**. The symbol $\in$ means "in" or "belongs to", and $\mathbb{R}$ denotes the set of real numbers (we shall define them in details in the next chapter). The symbol $\ni$ reads "such that", which is sometimes abbreviated as "s.t.". The statement means that there is at least one real number x that satisfies the equation that follows, which in this case is false. The reason is that no real number satisfies that equation $x^2 + 1 = 0$, since it implies that $x = \pm\sqrt{-1}$. We can, however, make the statement true by expanding the choice of x to include the set of complex numbers, $\mathbb{C}$:

$$\exists\, x \in \mathbb{C} \ni x^2 + 1 = 0.$$

In daily conversations, we do not use such a formal language. Instead of saying "there exists a bird such that it can swim", we say "some birds can swim".

Sometimes we want to assign a condition or property that is shared by all members of a group. In that case we need the **universal quantifier**. For example, we want to express the fact that the squared values of all real numbers is equal to or greater than zero. In logical symbols, it becomes

$$\forall\, x \in \mathbb{R}, \ x^2 \geq 0.$$

The symbol $\forall$ denotes "for all" or "for every". The statement requires that *all* members of the set $\mathbb{R}$ satisfy the condition that follows. In other words, if we can find even one real number x such that $x^2 < 0$, then the statement is false. For example, the statement

$$\forall\, x \in \mathbb{R}, \ x^2 > 0 \tag{1.1}$$

is false since when $x = 0$, $x^2 = 0$ instead of $x^2 > 0$. In general, if $p(x)$ is a statement that its truth value depends on a variable x, a universal statement is in the form

$$\forall x, \; p(x).$$

The above statement is true if every x in the context of our discussion makes $p(x)$ true. On the other hand, if there is one x such that $p(x)$ is false, the universal statement is false. It follows that the negation of a universal statement, $\sim (\forall x, \; p(x))$ is equivalent to

$$\exists x \ni \; \sim p(x).$$

For example, the negation of the statement in (1.1) is

$$\exists x \in \mathbb{R} \ni x^2 \le 0,$$

which reads "there exists a real number such that its square is less than or equal to zero".

The general form of an existential statement is

$$\exists x \ni p(x). \tag{1.2}$$

For the statement to be true, we need to find at least one x such that $p(x)$ is true. On the other hand, the statement is false if every x in question makes $p(x)$ a false statement. Therefore the negation of the existential statement in (1.2) is equivalent to

$$\forall x, \; \sim p(x).$$

When we write quantified statements involving more than one quantifier, their order is important because it can affect the meaning of the statement. Consider the following statements under the context of real numbers:

$$\exists x \ni \forall y, x > y, \tag{1.3}$$

and

$$\forall y, \exists x \ni x > y. \tag{1.4}$$

Statement (1.3) means that we can find a real number that is greater than any real number, which is false. On the other hand, statement (1.4) asserts that for any given real number, we can find a greater one.

1.4   Mathematical Proof

Consider the conditional statement "if the money supply in an economy increases faster than its growth in output, then inflation will follow". Economists spend a lot of time debating whether the statement is always true, sometimes true, or always false. In science, a "law" is a statement that has been conforming to empirical evidence. A general rule is that a scientific law is accepted as true until new observations show otherwise. The world of mathematics is more certain. A theorem derived from universally accepted axioms is always true. Such a statement is called a **tautology**. Therefore, in a sense, mathematics is the study of tautologies using logical rules. For example, the theorems in the book *Elements of Geometry*, written by the Greek mathematician Euclid in 300 B.C., are still true under the assumption of his postulates and axioms. In this section we discuss some useful techniques in mathematical proofs. Interested readers can consult books such as Beck and Geoghegan (2010) or Gerstein (2012) for more detailed discussions.

Before discussing the proof, we first define some common terms in mathematics. An **axiom** is a statement assumed to be true in a mathematical system. For example, in Euclid's *Elements*, one of the axioms (which he calls a common notion) is

Things that are equal to the same thing are also equal to one another.

This axiom is called "transitivity of equality", which means that if $a = b$ and $b = c$, then $a = c$. Axioms and definitions are the basic rules of a mathematical system They are analogous to the rules of a chess game—no question asked.[1]

A **proposition**, **conjecture** or **postulate** is a statement which contains ideas that can be true or false. One of the most famous conjectures is Fermat's Last Theorem, which states that no three positive integers a, b and c can satisfy the equation $a^n + b^n = c^n$ for any integer value of n greater than two. It was proposed by the French mathematician Pierre de Fermat in the seventeenth century. When it was finally proven by Andrew Wiles in 1995, it became a **theorem**, a logically proven true statement.

A **lemma** is a proven true statement that is useful in proving other results, while a **corollary** is a true statement that is a consequence of a theorem.

1.4.1   Direct Proof of Implications

Recall in Sect. 1.2 that a conditional statement is true when

- the hypothesis and the conclusion are both true, or
- the hypothesis is false, regardless the truth value of the conclusion.

---

[1]For a more detailed discussion of axioms and proof see Devlin (2002).

In the second case, the implication is true by definition, so there is nothing to prove. Therefore to prove that an implication is true directly, we need to assume that the hypothesis p is true and then "show" that the conclusion is also true. But what do we mean by "showing"? The following equivalent statement is helpful:

$$(p \Rightarrow q) \Leftrightarrow (p \Rightarrow p_1 \Rightarrow p_2 \Rightarrow \cdots \Rightarrow q_2 \Rightarrow q_1 \Rightarrow q).$$

That is, we can take a number of intermediate steps between p and q, and in each step we arrive at a new true statement ($p_1, p_2, \ldots, q_2, q_1$, etc.). We know that the intermediate conditional statements are true by invoking axioms, definitions and proven results.

For example, suppose we want to prove that if n is a positive integer, then $n^2 + 3n + 8$ is an even number. The key ingredients of this conditional statement are

- Hypothesis p: n is a positive integer,
- Conclusion q: $n^2 + 3n + 8$ is an even number.

For a direct proof, the first step is to assume that p is true, that is, n is a positive integer. It is not obvious at this stage how we can arrive at the conclusion q. One strategy is to do some backward thinking about what statement would imply q, that is, what is q_1? In this case the definition of an even number is helpful. An even number can be expressed as $2k$, where k is an integer. Therefore somehow we must be able to express $n^2 + 3n + 8$ as $2k$. Another useful property is that $(n + 1)(n + 2)$ is even since one of $n + 1$ or $n + 2$ is even and the product of an even number with any integer is even. Now we can write down the formal proof as follows.

Proof Suppose that n is a positive integer. Then $(n + 1)(n + 2) = 2m$ is even since one of $n + 1$ or $n + 2$ is even and the product of an even number with any integer is even. Now

$$\begin{aligned}
n^2 + 3n + 8 &= n^2 + 3n + 2 + 6 \\
&= (n + 1)(n + 2) + 6 \\
&= 2m + 2(3) \\
&= 2(m + 3) \\
&= 2k,
\end{aligned}$$

where m and $k = m + 3$ are integers. Therefore $n^2 + 3n + 8$ is even. □

You should be able to write down the intermediate statements in the above proof.

1.4.2   Proof by Contrapositives

Sometimes a direct proof of a conditional statement can be too complicated or too difficult. A very useful alternative method is by proving the contrapositive instead. The method depends on the following equivalent statements (see Exercise 10):

$$(p \Rightarrow q) \Leftrightarrow [(\sim q) \Rightarrow (\sim p)].$$

Therefore we proceed by assuming the negation of the conclusion is true and show that negation of the hypothesis is true.

Example 1.4 Suppose that $a < b$ and $f(x)$ is a real-valued function. Prove that if

$$\int_a^b f(x)\,dx \neq 0,$$

then there exists an x between a and b such that $f(x) \neq 0$.

Here the hypothesis and the conclusion are

- $p : \int_a^b f(x)\,dx \neq 0$,
- q: there exists an x between a and b such that $f(x) \neq 0$,

and we need to prove $p \Rightarrow q$. You can try the direct proof but the contrapositive of the statement provides an easy way out. The negations of p and q are

- $\sim p : \int_a^b f(x)\,dx = 0$,
- $\sim q$: For all x between a and b, $f(x) = 0$.

Now we proceed to show that $(\sim q) \Rightarrow (\sim p)$.

Proof Suppose that $a < b$ and $f(x)$ is a real-valued function. If $f(x) = 0$ for all x between a and b, then by the definition of integration $\int_a^b f(x)\,dx = 0$.   □

1.4.3   Proof by Contradiction

If both the direct method and the contrapositive fail to give a manageable proof of an implication, there is another alternative called proof by contradiction. The technique is based on the following equivalent statement.

$$(p \Rightarrow q) \Leftrightarrow [p \wedge (\sim q) \Rightarrow c],$$

where c is called a contradiction, defined as a statement that is always false. Therefore, instead of proving the implication directly, we assume that the hypothesis is true but the conclusion is false. Then we show that this implies a statement that is known to be false or contradicts our assumption.

Example 1.5 Prove that if x is rational and y is irrational, then $x + y$ is irrational.

Recall that a number x is rational if it can be expressed as ratio of two integers, that is, $x = a/b$, where a and b are integers and $b \neq 0$. Otherwise it is called irrational. Here we want to prove the conditional statement $p \Rightarrow q$ where

- p: x is rational and y is irrational,
- q: $x + y$ is irrational.

We can prove this by contradiction, that is, we show that $(p \wedge \sim q) \Rightarrow c$.

Proof Suppose that x is rational and y is irrational but $x+y$ is rational. By definition $x = a/b$ and $x + y = c/d$, where a, b, c, d are integers and $b, d \neq 0$. It follows that

$$y = \frac{c}{d} - \frac{a}{b} = \frac{bc - ad}{bd}.$$

Since $bc - ad$ and bd are integers and $bd \neq 0$, y is rational. This contradicts our assumption that y is irrational. Therefore $x + y$ is irrational. □

Another case of proof by contradiction is the simple equivalent statement

$$p \Leftrightarrow (\sim p \Rightarrow c).$$

That is, instead of showing directly the statement p is true, we can show that its negation implies a contradiction.

Example 1.6 The number $\sqrt{2}$ is irrational.

Proof Assume on the contrary that $\sqrt{2}$ is rational. Hence $\sqrt{2} = m/n$ for some integers m and n with no common factors. It follows that $m = \sqrt{2}n$ so that $m^2 = 2n^2$. This means that m^2 and therefore m are even integers. Writing $m = 2k$ for some integer k, we have $2n^2 = m^2 = 4k^2$, or $n^2 = 2k^2$. This implies that n is also even, which contradicts our assumption that m and n have no common factors. □

1.4.4 Proving Implications with Cases

Many conditional statements are in the form $p \Rightarrow (q \vee r)$. Instead of a direct proof, the following equivalent statement is very useful:

$$[p \Rightarrow (q \vee r)] \Leftrightarrow [(p \wedge (\sim q)) \Rightarrow r].$$

Of course the role of the statements q and r is symmetric. That is, we can show $(p \wedge (\sim r)) \Rightarrow q$ instead.

Example 1.7 Suppose that x is a real number. If $2x = x^2$, then $x = 0$ or $x = 2$.

Proof Suppose that $2x = x^2$ and $x \neq 0$. Dividing both side of the equation by x gives $x = 2$. $\qquad\qquad\square$

Another common conditional statement with case is in the form $(p \vee q) \Rightarrow r$. That is, the hypothesis can be divided into two cases. For example, an integer is either odd or even. Similarly, a real number x is either negative or $x \geq 0$. Or it is rational or irrational. A useful tool in proving this kind of statement is by using the following equivalent statement.

$$[(p \vee q) \Rightarrow r] \Leftrightarrow [(p \Rightarrow r) \wedge (q \Rightarrow r)].$$

That is, we show that both cases ensure the conclusion r. The following example employs this technique at the end of the proof.

Example 1.8 Show that there exists irrational numbers x and y such that x^y is rational.

Proof We have established in Example 1.6 above that $\sqrt{2}$ is irrational. But we observe that

$$\left(\sqrt{2}^{\sqrt{2}}\right)^{\sqrt{2}} = \sqrt{2}^{(\sqrt{2}\sqrt{2})} = \sqrt{2}^2 = 2,$$

which is rational. Now we have two cases. First, if $\sqrt{2}^{\sqrt{2}}$ is rational, then we let $x = y = \sqrt{2}$. Second, if $\sqrt{2}^{\sqrt{2}}$ is irrational, then we let $x = \sqrt{2}^{\sqrt{2}}$ and $y = \sqrt{2}$ as required. $\qquad\qquad\square$

Notice that in the above example we do not know whether $\sqrt{2}^{\sqrt{2}}$ is rational or irrational. But in both cases we can find two irrational numbers to give the results.

1.4.5   Proving Equivalent Statements

Recall that the general form of an equivalent statement is $p \Leftrightarrow q$, which is defined by $(p \Rightarrow q) \wedge (q \Rightarrow p)$. Therefore the proof of the statement involves two parts. First we prove that $p \Rightarrow q$. Then we proof its converse $q \Rightarrow p$. It is not necessary to use the same technique for both parts. For example, it is perfectly fine to prove the first implication by the direct method and the converse by using its contrapositive. The following example illustrates this point.

Example 1.9 Suppose that n is an integer. Prove that n is odd if and only if n^2 is odd.

Proof In the first part we show that n is odd implies that n^2 is odd. By definition of an odd number, $n = 2k + 1$, for some integer k. Therefore

$$n^2 = (2k + 1)^2 = 4k^2 + 4k + 1 = 2(2k^2 + 2k) + 1.$$

Since $2k^2 + 2k$ is also an integer, n^2 is an odd number. This completes the proof of the first part.

In the second part, we prove the converse by proving the contrapositive, which states that if n is even, then n^2 is even. So suppose that $n = 2k$ for some integer k. Then $n^2 = 4k^2 = 2(2k^2)$, which is even. □

1.4.6 Proving Quantitative Statements

A universal statement proclaims that *all* members of a set under discussion satisfy a certain property. For example, the statement "all birds can fly" means that if we pick an arbitrary animal classified as a bird, then it must be able to fly. Of course the statement is false as there are indeed flightless birds. Therefore, in proving a universal statement $\forall x, \ p(x)$, we need to pick an *arbitrary* member x in the set and show that $p(x)$ is true. To disprove a universal statement, we have to find a counter-example. That is, if there exists at least an x such that $\sim p(x)$, then the universal statement is false.

On the other hand, to prove an existential statement such as "some birds can fly", we need to find at least one bird in the sky. To disprove an existential statement $\exists x \ni p(x)$, we need to show that $p(x)$ is false for all x in question.

Example 1.10 Prove that for all real number x, there exists a number y such that $y > x$.

Proof Suppose that x is any arbitrary real number. Let $y = x + 1$. Then $y > x$ as required. □

Example 1.11 Prove that the square of all odd integers can be expressed as $8k + 1$ for some integer k.

Proof Let n be any odd integer. Then $n = 2m + 1$ for some integer m. Now

$$n^2 = (2m + 1)^2 = 4m^2 + 4m + 1 = 4m(m + 1) + 1.$$

Since either m or $m + 1$ is even, $4m(m + 1) = 8k$ where k is an integer. So we conclude that $n^2 = 8k + 1$.   □

Example 1.12 Prove that there exists a real number x such that for all y and z, if $y > z$, then $y > x + z$.

Proof Choose $x = 0$. Then for all real numbers y and z such that $y > z$, $y > 0 + z$.
                                                                                          □

Example 1.13 Suppose that $f(n) = 1 - 1/n$ for $n = 1, 2, 3, \ldots$. For all $\epsilon > 0$, show that there exists a positive integer N such that for all $n > N$,

$$|f(n) - 1| < \epsilon. \tag{1.5}$$

We need to take any arbitrary value of a positive ϵ and then find an N that all $n > N$ satisfy the inequality in (1.5). Now

$$|f(n) - 1| = |1 - 1/n - 1| = |-1/n| = 1/n.$$

Therefore we need to have $1/n < \epsilon$, or $n > 1/\epsilon$. Hence we can let $N = \lceil 1/\epsilon \rceil$, where $\lceil x \rceil$ is the ceiling function, defined as the smallest integer greater than the number x. For example, if $\epsilon = 0.07$, then $N = \lceil 1/0.07 \rceil = \lceil 14.29 \rceil = 15$. Here is the formal proof.

Proof Given any $\epsilon > 0$, let $N = \lceil 1/\epsilon \rceil$. Then for all $n > N$,

$$|f(n) - 1| = |1 - 1/n - 1| = |-1/n| = 1/n < 1/N \le \epsilon$$

as required.   □

It should be emphasized that finding a few examples is not a formal proof of a universal statement. Consider the statement "for all integers $n \ge 2$, $f(n) = 2^n - 1$ is a prime number". One can observe that $f(2) = 3$, $f(5) = 31$, $f(7) = 127$, etc. are prime numbers. There are, however, many integers n such that $f(n)$ are not primes. What we can claim is, "there exists integers $n \ge 2$ such that $f(n) = 2^n - 1$ are prime numbers", albeit not a very useful result.

Table 1.1 summarizes some useful tautologies used in proof. There is one more important tool called mathematical induction. But we defer the discussion until Chap. 2, where we explore a little bit more about natural numbers.

Table 1.1 Some useful
tautologies

(a)	$\sim (p \wedge q) \Leftrightarrow (\sim p) \vee (\sim q)$
(b)	$\sim (p \vee q) \Leftrightarrow (\sim p) \wedge (\sim q)$
(c)	$\sim [\forall x, \; p(x)] \Leftrightarrow [\exists x \ni \sim p(x)]$
(d)	$\sim [\exists x \ni p(x)] \Leftrightarrow [\forall x, \; \sim p(x)]$
(e)	$(p \Rightarrow q) \Leftrightarrow (p \Rightarrow p_1 \Rightarrow p_2 \Rightarrow \cdots \Rightarrow q_2 \Rightarrow q_1 \Rightarrow q)$
(f)	$(p \Rightarrow q) \Leftrightarrow [(\sim q) \Rightarrow (\sim p)]$
(g)	$(p \Rightarrow q) \Leftrightarrow [p \wedge (\sim q) \Rightarrow c]$
(h)	$p \Leftrightarrow (\sim p \Rightarrow c)$
(i)	$[p \Rightarrow (q \vee r)] \Leftrightarrow [(p \wedge (\sim q)) \Rightarrow r]$
(j)	$[(p \vee q) \Rightarrow r] \Leftrightarrow [(p \Rightarrow r) \wedge (q \Rightarrow r)]$
(k)	$[p \wedge (p \Rightarrow q)] \Leftrightarrow q$
(l)	$\sim (p \Rightarrow q) \Leftrightarrow [p \wedge (\sim q)]$
(m)	$[(p \vee q) \wedge (\sim q)] \Rightarrow p$
(n)	$[(p \Rightarrow q) \wedge (r \Rightarrow s) \wedge (p \vee r)] \Rightarrow (q \vee s)$
(o)	$[(p \wedge \sim q) \vee (q \wedge \sim p)] \Leftrightarrow [(p \vee q) \wedge (\sim p \vee \sim q)]$

1.5 Strategies in Mathematical Proof

Constructing a formal mathematical proof is a creative process, and there is no unique way to achieve that. In the chapters that follow you will read and work on many proofs, some are straightforward and some are tricky. Many times you will encounter a statement that you have no clue where to start. Here are some recommendations to follow and kick-start the thinking process.

First, decipher the logical structure of the statement. Is it a simple statement, a conditional statement, or a quantified statement? On a side note to your formal proof, write down what are given in the statement, and what are needed to be proven.

Second, recall and write down the definitions of every mathematical objects in the statement. A lot of simple proofs just need a few steps between the definitions. In making those steps, recall any known results that you or others have developed.

Sometimes an example or a diagram may help you to clarify some ideas. Keep in mind, however, that examples and diagrams are not formal proofs. One exception is that you are finding a counter-example to disprove a universal statement. For example, to disprove a universal statement such as $\forall x \in \mathbb{R}, \; p(x)$, all you need is to find one number x such that $p(x)$ is false. On the other hand, to prove such a statement, always begin your proof with "Let x be a real number. ...".

My final advice is that studying mathematics is not unlike learning a foreign language or participating in a sport. Practising is as important as reading and memorizing. Therefore working on the exercises at the end of each chapter is an important part of developing intuition, stimulating creativity and gaining experience. Over time you will be able to acquire that elusive quality called mathematical maturity. As one Olympic swimmer once said, "To be a good swimmer, you have to feel the water".

1.6   Exercises

1. Determine whether each of the following sentences is a statement. Provide a brief justification.
   (a) Some sentences can be labelled true and false.
   (b) All students who fail the first test should drop the course.
   (c) If Hillary Clinton becomes the president of the United States in 2016, she will be the best president in U.S. history.
   (d) The contrapositive of an implication is not equivalent to the implication.
   (e) When the central bank increases the supply of money, the interest rate will decrease and the quantity of money demanded will increase.

2. Determine whether each of the following sentences is a statement. Provide a brief justification.
   (a) All consumers have transitive preferences.
   (b) The inflation rate is negatively related to the unemployment rate.
   (c) All students should study mathematics.
   (d) This statement is not true.
   (e) This is not a statement.
   (f) If $f : X \rightarrow Y$ and $h : X \rightarrow Z$ are linear functions with kernel $f \subseteq$ kernel h, then there exists a linear function $g : f(X) \rightarrow Z$ such that $h = g \circ f$.

3. Determine whether each of the following statements is true or false. Provide a brief explanation.
   (a) All presidents of the United States, past or present, are men.
   (b) Given the function $f(x) = x^2$. If $f(x_1) = f(x_2)$, then $x_1 = x_2$.
   (c) If the equilibrium price of a commodity goes up, then the equilibrium quantity must decrease.
   (d) For all $\epsilon > 0$, there exist $\delta < 0$ such that $\delta \geq \epsilon$.
   (e) This statement is not true.

4. Consider the compound statement: If f is differentiable on an open interval (a, b) and if f assumes its maximum or minimum at a point $c \in (a, b)$, then $f'(c) = 0$.
   (a) Express the statement in four simple statements p, q, r and s with logical symbols $\Rightarrow, \forall, \exists, \ni, \vee, \wedge$, etc.
   (b) Find the negation of the statement in logical symbols.

5. If $\lim_{x \rightarrow a} f(x) \neq f(a)$, then f is not continuous at a. Also, f is differentiable at a implies that f is continuous at a. Prove that f is not differentiable at a whenever $\lim_{x \rightarrow a} f(x) \neq f(a)$.

6. Let p and q be logical statements. Prove the following De Morgan's laws:
   (a) $\sim (p \wedge q) \Leftrightarrow (\sim p) \vee (\sim q)$.
   (b) $\sim (p \vee q) \Leftrightarrow (\sim p) \wedge (\sim q)$.

7. Determine if an implication and its converse are equivalent.

8. Determine if an implication and its inverse are equivalent.

9. Consider the following statements[2]:

> Around the world, nearly every asset class is expensive by historical standards. ...The inverse of that is relatively low returns for investors.

Is the second statement the logical inverse of the first statement? Discuss.

10. Prove the following characterizations of an implication by constructing truth tables.
    (a) $(p \Rightarrow q) \Leftrightarrow [(\sim q) \Rightarrow (\sim p)]$.
    (b) $(p \Rightarrow q) \Leftrightarrow [(\sim p) \vee q]$.
    (c) $(p \Rightarrow q) \Leftrightarrow [p \wedge (\sim q) \Rightarrow c]$.

11. Use a truth table to show that each of the following statement is a tautology:
    (a) $[\sim q \wedge (p \Rightarrow q)] \Rightarrow \sim p$.
    (b) $\sim (p \Rightarrow q) \Leftrightarrow (p \wedge \sim q)$.

12. A political commentator asserts the statement $p \Rightarrow q$ where
    - p: For a number of reasons, country X cannot have democracy.
    - q: Democratic movement is suppressed by the government.
    (a) State the converse of the above statement in plain English.
    (b) State the contrapositive of the above statement in plain English.
    (c) If the government suddenly changes its view and promotes democracy in country X, what would the commentator say?

13. A prime number is a natural number that is divisible by 1 and itself only. Let $\mathbb{P}$ be the set of all prime numbers. Consider the statement "If n is a prime number, then $2n + 1$ is also a prime number".
    (a) Write the statement in logical and set symbols.
    (b) Write the contrapositive of the statement in plain English.
    (c) Prove or disprove the statement.

14. There exists a number x such that for all y, $f(x, y) \geq 0$.
    (a) Write the above statement in logical and set symbols.
    (b) Find the negation of the statement.

15. In each part below, the hypotheses are assumed to be true. Use the tautologies from Table 1.1 to establish the conclusion. Indicate which tautology you are using to justify each step.
    (a) Hypotheses: $r, \sim t, (r \wedge s) \Rightarrow t$
        Conclusion: $\sim s$
    (b) Hypotheses: $s \Rightarrow p, s \vee r, q \Rightarrow \sim r$
        Conclusion: $p \vee \sim q$
    (c) Hypotheses: $\sim s \Rightarrow t, t \Rightarrow r$
        Conclusion: $\sim r \Rightarrow s$
    (d) Hypotheses: $r \Rightarrow (\sim p \vee q)$
        Conclusion: $p \Rightarrow (q \vee \sim r)$

---

[2]Neil Irwin, "Welcome to the Everything Boom, or Maybe the Everything Bubble", *New York Times*, July 7, 2014.

16. Repeat the above exercise with the following statements.
   (a) Hypotheses: $r \Rightarrow \, \sim p, \; \sim r \Rightarrow q$
       Conclusion: $p \Rightarrow q$
   (b) Hypotheses: $q \Rightarrow \, \sim p, \; r \Rightarrow s, \; p \vee r$
       Conclusion: $\sim q \vee s$
   (c) Hypotheses: $p \Rightarrow s, s \Rightarrow q, \sim q$
       Conclusion: $\sim p$
   (d) Hypotheses: $\sim s \Rightarrow \, \sim p, (\sim q \wedge s) \Rightarrow r$
       Conclusion: $p \Rightarrow (q \vee r)$

17. Let n be a positive integer. Prove or disprove the following statements:
   (a) $2n$ is divisible by 4.
   (b) If n^2 is an odd number, then n is an odd number.

18. Let n be a positive integer. Prove or disprove the following two statements:
   (a) If n is divisible by 2 and by 5, then it is divisible by 10.
   (b) If n is divisible by 4 and by 6, then it is divisible by 24.
   (c) Can you prove the general statement? If n is divisible by l and k, which are distinct prime numbers. Then n is divisible by lk.

19. Let x be a real number. Prove or disprove the following statement:
       If $|x - 2| \leq 3$, then $-1 \leq x \leq 5$.

20. Let x be a rational number. Prove that if xy is an irrational number, then y is an irrational number.

21. Prove or disprove: If $5x$ is an irrational number, then x is also an irrational number.

22. Prove or disprove: For every positive integer n, the function given by

$$f(n) = n^2 + n + 13$$

gives a prime number.

23. Prove that if n is an odd integer, then $n^2 + 2n - 1$ is even.

24. Let m and n be integers. Prove that mn is even if and only if m is even or n is even.

25. Let x be a nonzero real number. If

$$x + \frac{1}{x} < 2,$$

then $x < 0$.
   (a) Prove the statement directly.
   (b) Repeat using the contrapositive.
   (c) Repeat using proof by contradiction.

26. Suppose that p is a prime number. Show that $\sqrt{p}$ is irrational.

References

Beck, M., & Geoghegan, R. (2010). *The art of proof*. New York: Springer Science+Business Media.

Devlin, K. (2002). Kurt Gödel—separating truth from proof in mathematics. *Science*, Volume 298, December 6 issue, 1899–1900.

Gerstein, L. J. (2012). *Introduction to mathematical structures and proofs*, Second Edition. New York: Springer Science+Business Media.

Sets and Relations

2

The language of sets is an important tool in defining mathematical objects. In this chapter we review and study set theory from an intuitive and heuristic approach. Readers who want to get the formal axiomatic treatment can consult books on set theory such as Devlin (1993).

2.1   Basic Notations

Although not a formal definition, a **set** can be understood to be a collection of well-defined objects. Each individual object is called an **element** or a **member** of the set. We express the idea that an object a is a member of a set A by $a \in A$, where the symbol $\in$ reads "is an element of", "belongs to", "is contained in" or simply "is in". Normally, but not always, we use lowercase letters to represent the elements and uppercase for sets. Instead of $\sim (a \in A)$, which means that a is not an element of A, we write $a \notin A$. There are several ways to describe a set. For examples,

$$A = \{\text{Sonata}, 7, \text{tree}, \$2.75, \text{utility}, \pi, \text{C++}, \text{entropy}\},$$

$$B = \{1, 2, 3, \ldots, 9\},$$

$$C = \{x : x \text{ is a day of the week}\}.$$

We describe the set A by listing all of the elements between the braces { }. Elements are separated by commas. Although the elements seem to be unrelated objects, nevertheless their meanings are well defined. The set B contains the integers from 1 to 9. The elements represented by the ellipsis "$\ldots$" are to be understood in the context. The set C contains any element x which satisfies the property described after the colon (:). Therefore it can also be listed as

$$C = \{\text{Sunday}, \text{Monday}, \text{Tuesday}, \text{Wednesday}, \text{Thursday}, \text{Friday}, \text{Saturday}\}.$$

© Springer Nature Switzerland AG 2019
K. Yu, *Mathematical Economics*, Springer Texts in Business and Economics,
https://doi.org/10.1007/978-3-030-27289-0_2

In general, if $p(x)$ is a statement whose truth value depends on the variable x, then the set $\{x : p(x)\}$ contains all elements x such that $p(x)$ is true.

Being a general idea, a set can contain elements which are themselves sets. For example, $S = \{A, B, C\}$, where A, B and C are sets defined above. A set of sets is often called a **class** or a **collection**.

Given two sets A and B, we say A is a **subset** of B, or $A \subseteq B$, if $x \in A$ implies that $x \in B$. That is, all the elements belongs to A are also in B. Two sets are **equal** if they contain the same elements. In other words,

$$(A = B) \Leftrightarrow [(A \subseteq B) \wedge (B \subseteq A)].$$

The set A is said to be a **proper subset** of B if $A \subseteq B$ but $A \neq B$. Therefore, there exists some elements in B that are not in A. In this case we write $A \subset B$.

Some important sets in mathematics are listed below:

Natural Numbers $\mathbb{N} = \{1, 2, 3, \dots\}$
Integers $\mathbb{Z} = \{\dots, -3, -2, -1, 0, 1, 2, 3, \dots\}$
Rational Numbers $\mathbb{Q} = \{n/m : n, m \in \mathbb{Z}, m \neq 0\}$
Real Numbers $\mathbb{R} = (-\infty, \infty)$
Complex Numbers $\mathbb{C} = \{a + bi : a, b \in \mathbb{R}, i = \sqrt{-1}\}$

The set of natural numbers $\mathbb{N}$ and the number zero can be formally constructed with set theory using an idea called ordinal. Alternatively, we can accept the fact that many animals, including humans, have the innate ability to count and to add.[1] Then given a natural number n, we ask the question whether there exists a number x such that $n + x = 0$. This gives rise to the idea of negative integers. Next, given two integers n and m, we ask the question if there exists a number x such that $n = mx$. The results are the rational numbers. We shall not give a precise definition of the set of real numbers here. Interested readers can find the process of constructing real numbers from the rational numbers in many books on analysis.[2] The set $\mathbb{R}$ contains all rational numbers in $\mathbb{Q}$ and all irrational numbers such as $\sqrt{2}, \sqrt{5}, \pi$, etc. Geometrically, it contains all points on the real line. You should convince yourself that

$$\mathbb{N} \subset \mathbb{Z} \subset \mathbb{Q} \subset \mathbb{R} \subset \mathbb{C}.$$

In a mathematical system, it is often desirable to define a **universal set** U which contains all the elements under discussion. For example, in number theory, the universal set is the set of integers $U = \mathbb{Z}$. In real analysis, we often refer to the set of real numbers $U = \mathbb{R}$. By definition, any set under discussion is a subset of U. On the other hand, it is convenient to define a set with no element, which is called

---

[1] See Holt (2008) or Economist (2008).

[2] See, for example, Rudin (1976). The colourful description of the number systems in Strogatz (2010) is entertaining.

the **empty set** $\varnothing$. The empty set is a subset of any set. Given a set S, the **power set** of S, denoted by $\mathcal{P}(S)$, is the collection of all subsets of S.

Example 2.1 Let $S = \{a, b, c\}$. Then

$$\mathcal{P}(S) = \{\varnothing, \{a\}, \{b\}, \{c\}, \{a, b\}, \{a, c\}, \{b, c\}, \{a, b, c\}\}.$$

The **cardinality** of a set S is defined as the number of distinct elements in the set and is denoted by $|S|$. For example, in the sets defined above,

$$|A| = 8, \quad |B| = 9, \quad |C| = 7, \quad |S| = 3, \quad |\mathcal{P}(S)| = 8.$$

These are called **finite sets** because they contain a finite number of elements. Sets whose elements can be listed one by one are called **countable** or **denumerable**. For example, $\mathbb{N}$, $\mathbb{Z}$ and $\mathbb{Q}$ are countable sets. Also, any finite set is countable. On the other hand, $\mathbb{R}$ and $\mathbb{C}$ are uncountable.

Theorem 2.1 *Let S be a finite set with cardinality $|S| = n$. Then $|\mathcal{P}(S)| = 2^n$.*

Proof We shall give an informal proof here. In each subset of S, an element is either in the set or not in the set. Therefore there are two choices for each element. The total number of combinations is therefore

$$2 \times 2 \times \cdots \times 2 = 2^n$$

for all the n elements of S. $\qquad\qquad\qquad\qquad\qquad\qquad\qquad\qquad\qquad\quad\square$

Since the natural numbers are unbounded, there is an infinite number of them. This is also true for $\mathbb{Z}$, $\mathbb{Q}$, $\mathbb{R}$ and $\mathbb{C}$. It turns out, however, that there are different sizes of infinities. The cardinality of the set of natural numbers is denoted by $|\mathbb{N}| = \aleph_0$ (pronounced as aleph naught), while the cardinality of the set of real number is denoted by $|\mathbb{R}| = c$. Some facts on cardinality are listed as follows.[3]

1. For any set S, $|S| < |\mathcal{P}(S)|$.
2. $|\mathbb{N}| = |\mathbb{Z}| = |\mathbb{Q}| = \aleph_0$.
3. $|\mathcal{P}(\mathbb{N})| = c$. Therefore $\aleph_0 = |\mathbb{N}| < |\mathbb{R}| = c$.

We should be cautious that **infinity**, denoted by ∞, is a mathematical symbol of unboundedness. For example, an interval on the real line

$$(a, \infty) = \{x \in \mathbb{R} : x > a\}$$

---

[3]For a formal treatment, see for example Gerstein (2012). A brief discussion can be found in Matso (2007).

represents the set of all real numbers greater than a. We do not generally treat ∞ as a number. More importantly, infinity is implied by the logical structure of the number systems. As a concept in science, however, it is unreachable, unobservable and untestable. This distinction between mathematics and reality is emphasized by Ellis (2012) in what he calls Hilbert's Golden Rule:

> If infinities occur as a core part of your explanatory model, it is not science. (p. 27)

As a consequence, we use infinity in scientific and economic models just to indicate that a variable under consideration is a large number.

2.2   Set Operations

Given two or more sets, we frequently merge, subtract, separate, concentrate and negate them. In this section we define the formal rules of manipulating sets.

Suppose that A and B are subsets of a universal set U. The **intersection** of A and B, denoted by $A \cap B$, is a set containing elements both in A and B. That is,

$$A \cap B = \{x : (x \in A) \wedge (x \in B)\}.$$

The **union** of A and B, denoted by $A \cup B$, is a set containing element either in A or in B, or both:

$$A \cup B = \{x : (x \in A) \vee (x \in B)\}.$$

It is a trivial exercise to show that $(A \cap B) \subseteq (A \cup B)$. Two sets A and B are said to be **disjoint** if they have no element in common. That is, $A \cap B = \varnothing$.

The **difference** of two sets, denoted by $A \setminus B$, is the set which contains the elements in A but not in B:

$$A \setminus B = \{x : (x \in A) \wedge (x \notin B)\}.$$

The set A^c is called the **complement** of A, which contains all the element not in A. In other words, $A^c = U \setminus A$.[4]

Example 2.2 Prove the first De Morgan's law of sets: $(A \cup B)^c = A^c \cap B^c$.

Proof To show equality, we need to show that $(A \cup B)^c \subseteq A^c \cap B^c$ and $A^c \cap B^c \subseteq (A \cup B)^c$. So first suppose that $x \in (A \cup B)^c$. This means that $x \notin (A \cup B)$, which implies that $x \notin A$ and $x \notin B$. By definition of the complement this means $x \in A^c \cap B^c$. Therefore $(A \cup B)^c \subseteq A^c \cap B^c$.

---

[4]Some writers denote the complement of A by A', while others prefer $\bar{A}$.

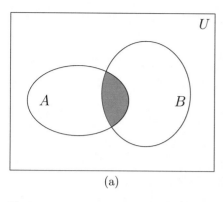 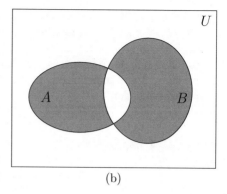

Fig. 2.1 Venn diagrams. (**a**) $A \cap B$. (**b**) $A \triangle B$

Conversely, let $x \in A^c \cap B^c$. Then x is not in A and also not in B, which can be written as $x \notin A \cup B$. Again by definition $x \in (A \cup B)^c$. Hence $A^c \cap B^c \subseteq (A \cup B)^c$ as required. □

Before attempting a formal proof, it is sometimes helpful to illustrate a set operation with a diagram. Often, in a Venn diagram, we represent the universal set with a rectangle, in which all subsets are drawn inside. Figure 2.1a depicts the intersection of two sets A and B, while Fig. 2.1b illustrates their **symmetric difference**, which is defined as

$$A \triangle B = (A \setminus B) \cup (B \setminus A).$$

Let $A_1, A_2, A_3, \ldots$ be a collection of sets. Suppose we want to form the union of all sets in the collection. We can denote the operation as

$$\bigcup_{n \in \mathbb{N}} A_n = A_1 \cup A_2 \cup A_3 \cup \cdots .$$

Similarly, the intersection of the sets can be denoted by

$$\bigcap_{n \in \mathbb{N}} A_n = A_1 \cap A_2 \cap A_3 \cap \cdots .$$

These are examples of **indexed collection of sets**. The set $\mathbb{N}$ is called an **index set**. The index set can be finite, infinitely countable or uncountable.

Example 2.3 Let $A_n = (-1/n, 1/n), n \in \mathbb{N}$. Then

$$\bigcup_{n \in \mathbb{N}} A_n = (-1, 1), \qquad \bigcap_{n \in \mathbb{N}} A_n = \{0\}.$$

Let S be a set and I an index set. We say that an indexed collection of subsets of S, $\{A_\alpha \subseteq S : \alpha \in I\}$, is a **partition** of S if

1. for all $\alpha \in I$, $A_\alpha \neq \varnothing$, that is, the subsets are nonempty;
2. for all $\alpha, \beta \in I$ such that $A_\alpha \neq A_\beta$, $A_\alpha \cap A_\beta = \varnothing$, that is, distinct subsets are pairwise disjoint;
3. $\bigcup_{\alpha \in I} A_\alpha = S$.

Example 2.4 Suppose that

$$S = \{x : x \text{ is a day of the year}\},$$

$$I = \{\alpha : \alpha \text{ is a day of the week}\}.$$

Then S is partitioned into seven subsets:

$$A_1 = \{x \in S : x \text{ is a Monday}\},$$

$$A_2 = \{x \in S : x \text{ is a Tuesday}\},$$

$$\vdots$$

$$A_7 = \{x \in S : x \text{ is a Sunday}\}.$$

Example 2.5 The indexed collection $\{A_n = (n, n+1] : n \in \mathbb{Z}\}$ forms a partition of the set $\mathbb{R}$.

In defining a set, the order of the elements in the list is not important. For example, $\{a, b\} = \{b, a\}$. In some situations, however, we want to define an **ordered set**. For example, an **ordered pair** of two elements a and b is defined as

$$(a, b) = \{\{a\}, \{a, b\}\},$$

whereas

$$(b, a) = \{\{b\}, \{a, b\}\}.$$

It is obvious that $(a, b) \neq (b, a)$. Therefore, given two ordered pairs, $(a, b) = (c, d)$ if and only if $a = c$ and $b = d$. We are now ready to define the **Cartesian product** of two sets A and B:

$$A \times B = \{(a, b) : a \in A, b \in B\}.$$

In general, $A \times B \neq B \times A$ unless $A = B$. The concept can be expanded into multi-dimensional product sets. Let $A_1, A_2, \ldots, A_n$. Then

$$A_1 \times A_2 \times \cdots \times A_n = \{(a_1, a_2, \ldots, a_n) : a_i \in A_i, i = 1, 2, \ldots, n\}.$$

The elements in the product set, $(a_1, a_2, \ldots, a_n)$, are called n-**tuples**. If $A_1 = A_2 = \cdots = A_n = A$, we write the product set as A^n. The following are examples of product sets.

Example 2.6 Let $A = \{a, b\}$ and $B = \{1, 2, 3\}$. Then

$$A \times B = \{(a, 1), (a, 2), (a, 3), (b, 1), (b, 2), (b, 3)\}.$$

Generally, if A and B are finite sets with $|A| = n$ and $|B| = m$, then $|A \times B| = nm$.

Example 2.7 The set of complex numbers $\mathbb{C}$ is the product set $\mathbb{R}^2$ with two operations, namely addition and multiplication. For any (a, b) and (c, d) in $\mathbb{C}$,

$$(a, b) + (c, d) = (a + c, b + d),$$
$$(a, b) \times (c, d) = (ac - bd, ad + bc).$$

The first number a in (a, b) is call the real part and the second number b is the imaginary part.

Example 2.8 Define $\mathbb{R}_+ = \{x \in \mathbb{R} : x \geq 0\}$, that is, the set of nonnegative real numbers. In consumer analysis, $\mathbf{x} = (x_1, x_2, \ldots, x_n) \in \mathbb{R}^n_+$ is called a consumption bundle with n goods and services. Each component x_i in the n-tuple represents the quantity of product i in the bundle. Similarly, define $\mathbb{R}_{++} = \{x \in \mathbb{R} : x > 0\}$, the set of positive real numbers. Then the corresponding prices of the goods and services in a bundle is the n-tuple $\mathbf{p} = (p_1, p_2, \ldots, p_n) \in \mathbb{R}^n_{++}$. Total expenditure is therefore the dot product of the two n-tuples, defined by

$$\mathbf{p} \cdot \mathbf{x} = p_1 x_1 + p_2 x_2 + \cdots + p_n x_n.$$

Example 2.9 In production theory, an n-tuple $\mathbf{y} = (y_1, y_2, \ldots, y_n) \in Y \subset \mathbb{R}^n$ represents a net output vector of a firm. A negative value for a component y_i implies that good or service i is an input factor in the production process, otherwise it is an output good or service. The production technology chosen by the firm determines the shape of the production set Y. The dot product of prices and the net output vector is the profit of the firm:

$$\pi = \mathbf{p} \cdot \mathbf{y} = p_1 y_1 + p_2 y_2 + \cdots + p_n y_n.$$

2.3 Binary Relations

Given two sets A and B, a **relation** R between A and B is a subset of $A \times B$. That is, $R \subseteq A \times B$. If $A = B$, we say that R is a relation on A. Instead of $(a, b) \in R$, we often write $a \, R \, b$, or a is related to b by R.

Example 2.10 Let N be a set of nuts and B be a set of bolts. We can define $(n, b) \in R$ if the size of nut n fits the size of bolt b.

Example 2.11 Let A be the set of people. A relation R on A can be defined as "is the mother of". The statement "Mary is the mother of James." can be expressed as (Mary, James) $\in R$.

Example 2.12 A relation R on $\mathbb{N}$ can be defined as "is a prime factor of". For examples, $(5, 20)$ and $(7, 56)$ are in R but $(3, 13) \notin R$.

Example 2.13 Let $m, n \in \mathbb{Z}$. We say m divides n if there exists $k \in \mathbb{Z}$ such that $n = km$. Instead of $m \, R \, n$, we have a special symbol for the relation, $m|n$. For examples, $4|32$, $7|49$, $12|132$ but $3 \nmid 23$.

Example 2.14 Let $\mathbb{R}^n_+$ be the consumption set of n goods and services of a consumer. We can define a preference relation $\succsim$ on $\mathbb{R}^n_+$. For consumption bundles $a, b \in \mathbb{R}^n_+$, $a \succsim b$ means that bundle a is at least as good as bundle b.

Given a relation R between A and B, we call the set A the **domain** of R and B the **codomain** of R. The **range** of R is defined as the set of elements in the codomain that is in the relation. That is,

$$\text{Range } R = \{b \in B : (a, b) \in R \text{ for some } a \in A\}.$$

By definition the range is a subset of the codomain of a relation. We can also define an **inverse relation** R^{-1} from B to A such that

$$b \, R^{-1} \, a \Leftrightarrow a \, R \, b,$$

or formally,

$$R^{-1} = \{(b, a) \in B \times A : (a, b) \in R\}.$$

Notice that the domain of R^{-1} is B and the codomain is A.

There are some properties that we frequently encounter in a binary relation R on a set A. In the following definitions, a, b and c are arbitrary elements in A.

Complete: $(a \, R \, b) \vee (b \, R \, a)$.
Reflexive: $a \, R \, a$.
Symmetric: $(a \, R \, b) \Rightarrow (b \, R \, a)$.
Transitive: $[(a \, R \, b) \wedge (b \, R \, c)] \Rightarrow (a \, R \, c)$.
Circular: $[(a \, R \, b) \wedge (b \, R \, c)] \Rightarrow (c \, R \, a)$
Asymmetric: $(a \, R \, b) \Rightarrow \sim (b \, R \, a)$.
Antisymmetric: $[(a \, R \, b) \wedge (b \, R \, a)] \Rightarrow (a = b)$.

Example 2.15 Consider the relation "divides" on $\mathbb{Z}$ from Example 2.13. As the example shows, not any pair of numbers divides each other so the relation is not complete. It is true that for all $n \in \mathbb{Z}, n|n$. Therefore the relation is reflexive. If $m|n$, it is not necessary that $n|m$ unless $m = n$. It follows that the relation is not symmetric but antisymmetric. If $m|n$ and $n|p$, then by definition there exists $k, l \in \mathbb{Z}$ such that $n = km$ and $p = ln$. Hence $p = (lk)m$ so that $m|p$ but in general $p \nmid m$. We conclude that $|$ is transitive but not circular. Finally, reflexivity implies that the relation is not asymmetric.

A relation R on a set S is called a **partial order** if R is reflexive, transitive and antisymmetric. The set S is called a **partial-ordered set**, or **poset**. For example, the relation $m|n$ on $\mathbb{Z}$, as shown in Example 2.15, is a partial order.

Example 2.16 Let S be a set. The relation $\subseteq$ on the power set of S, $\mathcal{P}(S)$, is a partial order.

A complete partial order is called a **linear order**. For example, the relation $\geq$ on $\mathbb{R}$ is a linear order. Notice that completeness implies reflexivity (but the converse is not true), so the latter assumption is redundant.

Another important class of relations in mathematics is the **equivalence relation**, which is defined as a relation R on a set S that is reflexive, transitive and symmetric. The relation "$=$" on the set of real numbers $\mathbb{R}$ is an example of equivalence relation. An equivalence relation is often denoted by an ordered pair $(S, \sim)$, where we use the symbol $\sim$ in place of R. For any $a \in S$, we define the **equivalence class** of a as

$$\sim (a) = \{x \in S : x \sim a\}.$$

Example 2.17 In Example 2.7 we express a complex number as an ordered pair in $\mathbb{R}^2$. For any complex number $z = (a, b)$, we define its absolute value as $|z| = \sqrt{a^2 + b^2}$. Then the relation $z \sim w$, which means that $|z| = |w|$, is an equivalence relation. The equivalence class $\sim (z)$ contains all the numbers that have absolute values equal to $|z|$. For example, the conjugate of z, $\bar{z} = (a, -b)$, belongs to $\sim (z)$.

The following theorem shows that the collection of the distinct equivalence classes in fact forms a partition of the set.

Theorem 2.2 *Suppose that* $(S, \sim)$ *is an equivalence relation where S is nonempty. Then the collection of equivalence classes* $\{\sim(a) : a \in S\}$ *is a partition of S.*

Proof By reflexivity, for every $a, b \in S$, $\sim(a)$ and $\sim(b)$ are nonempty. Suppose that $\sim(a) \cap \sim(b) \neq \varnothing$. Then there exists an $x \in \sim(a)$ such that x is also in $\sim(b)$. This means that $a \sim x$ and $x \sim b$, and by transitivity, $a \sim b$. It follows that $\sim(a) = \sim(b)$. Finally, $a \in \sim(a)$ for all $a \in S$. Therefore $\cup_{a \in S} \sim(a) = S$. We conclude that $\{\sim(a) : a \in S\}$ is a partition of S. $\square$

We can also show that the converse of Theorem 2.2 is also true.

Theorem 2.3 *Suppose that I is an index set and* $\{A_\alpha \subseteq S : \alpha \in I\}$, *is a partition of S. Then there exists an equivalence relation* $(S, \sim)$ *such that for each* $a \in S$, *the equivalence class* $\sim(a) = A_\alpha$ *for some* $\alpha \in I$.

Proof Define a relation R on S such that,

$$R = \{(a, b) \in S^2 : a, b \in A_\alpha \text{ for some } \alpha \in I\}.$$

Obviously for all $a \in S$, $a R a$ so that R is reflexive. For any $(a, b) \in R$, we have $(b, a) \in R$ by definition. Therefore R is symmetric. Suppose that (a, b) and (b, c) are in R. Then $a, b \in A_\alpha$ for some $\alpha \in I$. Also, $b, c \in A_\beta$ for some $\beta \in I$. Since b is in both A_α and A_β, $A_\alpha = A_\beta$. Therefore $(a, c) \in R$ so that R is transitive. $\square$

Therefore, a partition on a set S induces an equivalence relation on S and vice versa. They are the two sides of the same coin.

Example 2.18 In Example 2.4 we partition the days in a year into seven subsets, $A_1, \ldots, A_7$, according to the day of the week. We can define an equivalence relation as $a \sim b$ if a and b belong to the same day of the week. If a is a Monday, then $\sim(a) = A_1$. If b is a Tuesday, then $\sim(b) = A_2$, and so on.

In the partial ordering and equivalence relations we require them to be antisymmetric and symmetric respectively. A more general relation is a **preorder**, which is defined as reflexive and transitive. It is sometimes called a **preference relation** or simply an **ordering** and is denoted by $(S, \succsim)$. In what follows we shall call it an ordering. If $a \succsim b$ in an ordering, we say that a **succeeds** b, or, alternatively, b **precedes** a.

Example 2.19 In Example 2.14 we define a consumer preference relation $\succsim$ on the consumption set $\mathbb{R}^n_+$. In consumer theory we generally assume that the relation is complete (and hence reflexive) and transitive. Therefore mathematically $\succsim$ is an ordering. With this basic preference relation we also define two more relations. Suppose that $a, b \in \mathbb{R}^n_+$:

1. If $a \succsim b$ and $b \succsim a$, then we say bundle a is indifferent to bundle b, which is denoted by $a \sim b$. In fact we impose symmetry on the ordering and turn it into an equivalence relation. The equivalence class of a bundle, $\sim (a)$, is called the indifference set of a in economics.
2. If $a \succsim b$ and $b \nsim a$, then we say bundle a is strictly preferred to bundle b, which is denoted by $a \succ b$. Mathematically, we drop reflexivity and add asymmetry to the ordering and turn it into a **strict ordering**.

With the above notations on an ordering $(S, \succsim)$, we can define different types of **intervals**:

$$[a, b] = \{x \in S : (x \succsim a) \wedge (b \succsim x)\},$$

$$(a, b) = \{x \in S : (x \succ a) \wedge (b \succ x)\},$$

$$(a, b] = \{x \in S : (x \succ a) \wedge (b \succsim x)\},$$

$$[a, b) = \{x \in S : (x \succsim a) \wedge (b \succ x)\}.$$

The first definition is called a **closed interval**, the second an **open interval**. The last two definitions are sometimes called **half-open intervals**. Unbounded intervals are expressed as contour sets. In particular, for any $a \in S$, the **upper contour set** of a is defined as

$$\succsim (a) = \{x \in S : x \succsim a\},$$

whereas the **lower contour set** of a is

$$\precsim (a) = \{x \in S : a \succsim x\}.$$

You should convince yourself that $\succsim (a) \cap \precsim (a) = \sim (a)$.

An element m of an ordering $(S, \succsim)$ is called a **maximal element** if for all $x \in S$, $x \nsucc m$. That is, there is no element in S strictly succeeding m. An alternative definition is that for all $x \in S$, if $x \succsim m$, then $x \sim m$. An element g is called the **greatest element** if $g \succsim x$ for all $x \in S$, that is, g succeeds all elements in S. Notice that in the definition of a maximal element, we do not assume that the ordering is complete. Therefore there can be elements in S which are not related to m. In the definition of the greatest element, however, g succeeds all other elements in S so that it is assumed to relate to all elements in S. The greatest element of a set A must be a maximal element, but the converse is not true. The definitions of a **minimal element** and the **least element** are left as an exercise.

Example 2.20 The countries which constitute the United Nations are supposed to be equal. Therefore the head of states of the countries are unrelated to each other in their lines of authority so that all of them are maximal elements at the U.N. Within its own government, however, each head of state is the greatest element in political ranking.

Example 2.21 In consumer theory we usually assume that the consumption set $\mathbb{R}_+^n$ has no greatest element under the preference relation $\succsim$ since the set $\mathbb{R}_+^n$ is unbounded. Nevertheless, the choice set of the consumer is always a bounded subset of $\mathbb{R}_+^n$ because of financial and other constraints. A maximal element in the choice set is called an optimal bundle.

Let A be a subset of an ordered set S. An element $s \in S$ is an **upper bound** of A if for all $x \in A$, $s \succsim x$. If an upper bound exists we call A bounded above. Let B be the set of all upper bounds of A, that is,

$$B = \{s \in S : s \succsim x, x \in A\}.$$

Then the least element or a minimal element in B is called the **least upper bound** or **supremum** of A, which is denoted by sup A. Notice that sup A may or may not be an element of A. If it does, sup A is also a maximal element of A, often denoted by max A.

Similarly, an element $e \in S$ is a **lower bound** of A if for all $x \in A$, $x \succsim e$. If a lower bound exists we call A bounded below. Let L be the set of all lower bounds of A. Then the greatest element or a maximal element in L is called the **greatest lower bound** or **infimum** of A, which is denoted by inf A. Again inf A may or may not be an element of A. If it does, it is also a minimal element of A, often denoted by min A. The set A is said to be **bounded** if it is bounded above and bounded below.

Theorem 2.4 *Let S be a linear order and suppose that A is a nonempty bounded subset of S. Then* sup A *is unique.*

Proof Suppose that x and y are both sup A. Then by definition $x \succsim y$ and $y \succsim x$. Since a linear order is antisymmetric, $x = y$. □

The uniqueness of inf A of a linear order is left as an exercise. Consequently, every nonempty bounded subset of $\mathbb{R}$ has a unique supremum and a unique infimum.

Example 2.22 Let $A = \{1/n : n \in \mathbb{N}\} \subseteq \mathbb{R}$. Then sup $A = 1$ and inf $A = 0$. Notice that inf A does not belongs to A.

2.4   Well Ordering and Mathematical Induction

A relation R on a set S is a **well ordering** if it is

1. complete,
2. transitive,

3. antisymmetric,
4. every nonempty subset of S has a least element.

That is, S is a linearly ordered set with a least element for every subset. The sets $\mathbb{Z}$, $\mathbb{Q}$ and $\mathbb{R}$ are not well ordered with respect to the linear order $\geq$. For example, the set of negative integers, $\{\ldots, -3, -2, -1\}$, does not have a least element. The set $\{1/n : n \in \mathbb{N}\}$, as a subset of either $\mathbb{Q}$ or $\mathbb{R}$, do not have a least element as well. It turns out that the set of natural numbers $\mathbb{N}$ satisfies the requirements. We generally regard this property as an axiom called the **Well-Ordering Principle**:

Axiom 2.1 *If A is a nonempty subset of $\mathbb{N}$, then there exists a number $l \in A$ such that for all $k \in A$, $k \geq l$.*

With the well-ordered principle, we can prove a useful tool in proof called the **Principle of Mathematical Induction**. It involves proving statements that depend on the set of natural numbers, which economists frequently encounter in dynamic models.

Theorem 2.5 *Let $p(n)$ be a statement whose truth value depends on $n \in \mathbb{N}$. Suppose that*

1. $p(1)$ is true,
2. for all $k \in \mathbb{N}$, $p(k)$ implies $p(k + 1)$.

Then $p(n)$ is true for all $n \in \mathbb{N}$.

Proof Suppose that $p(1)$ is true and $p(k)$ implies $p(k + 1)$ for any integer $k > 1$. Suppose on the contrary that there exists m such that $p(m)$ is false. Let the nonempty set

$$A = \{m \in \mathbb{N} : p(m) \text{ is false}\}.$$

By Axiom 2.1, there exists a number $l \in A$ such that $m \geq l$ for all $m \in A$. Since $p(1)$ is true, l must be greater than 1. Then $l - 1 \notin A$ since l is the least element in A. This means that $p(l - 1)$ is true. By hypothesis $p(l)$ is true, contradicting the fact that $l \in A$. We conclude that $p(n)$ is true for all $n \in \mathbb{N}$. $\square$

To apply the principle of mathematical induction to prove that a statement $p(n)$ is true for $n = 1, 2, 3, \ldots$, we need to show that $p(1)$ is true and for any $k \in \mathbb{N}$, $p(k)$ implies $p(k + 1)$.

Example 2.23 Show that for all $n \in \mathbb{N}$,

$$\sum_{i=1}^{n} i = \frac{n(n+1)}{2}.$$

Proof For $n = 1$, $\sum_{i=1}^{n} i = n(n+1)/2 = 1$ so that $p(1)$ is true. Next, assume that $p(k)$ is true for any k, that is,

$$\sum_{i=1}^{k} i = \frac{k(k+1)}{2}.$$

Then

$$\sum_{i=1}^{k+1} i = \sum_{i=1}^{k} i + (k+1)$$
$$= \frac{k(k+1)}{2} + (k+1)$$
$$= \frac{(k+1)(k+2)}{2}.$$

This shows that $p(k+1)$ is true and so by induction $p(n)$ is true for all $n \in \mathbb{N}$. □

For any number $m \in \mathbb{N}$, the set $S = \{n \in \mathbb{N} : n \geq m\}$ is a well-ordered set. The principle of mathematical induction can be modified to prove statements such as "for all $n \in S$, $p(n)$ is true". In this case, we need to show that $p(m)$ is true and for all $k > m$, $p(k)$ implies $p(k+1)$.

Example 2.24 Show that $2^n \geq n^2$ for all integers $n \geq 4$.

Proof For $n = 4$, $2^4 = 16 = 4^2$ and so $p(4)$ is true. Now assume that $2^k \geq k^2$ for any $k > 4$. Then

$$2^{k+1} = 2(2^k)$$
$$\geq 2(k^2)$$
$$\geq k^2 + 5k \qquad \text{(since } k > 4)$$
$$= k^2 + 2k + 3k$$
$$\geq k^2 + 2k + 1$$
$$= (k+1)^2.$$

Therefore by induction $p(n)$ is true for all $k \geq 4$. □

2.5   Functions

Let A and B be nonempty sets. A **function** f from A to B, denoted by $f : A \rightarrow B$, is a relation such that each element in A is related to at most one element in B. Instead of (a, b) or $a R b$, we write $f(a) = b$. Therefore, if $f(a) = b$ and $f(a) = c$, then $b = c$ in B. A function is often called a **mapping** or a **transformation**. As defined in a relation before, the set A is the **domain** of f, and B is the **codomain**. For any $C \subseteq A$, the **image** of C, denoted by $f(C)$, is a subset of B such that

$$f(C) = \{b \in B : f(a) = b, a \in C\}.$$

The image of the domain of f is called the **range** of f, that is, range $f = f(A)$. It is clear that the range of a function is a subset of the codomain. For any $D \subseteq B$, the **pre-image** of D, denoted by $f^{-1}(D)$, is a subset in A such that

$$f^{-1}(D) = \{a \in A : f(a) \in D\}.$$

For any $b \in B$, we often call the pre-image $f^{-1}(b)$ the **contour** of b.

Sometimes we want to express the function as a subset of the product set $A \times B$. In this case we call it the **graph** of f:

$$\text{graph } f = \{(a, b) \in A \times B : f(a) = b\}.$$

Example 2.25 Let $A = \{1, 2, 3, 4\}$ and $B = \{a, b, c\}$. Define $f : A \rightarrow B$ by its graph as

$$\text{graph } f = \{(1, a), (2, c), (4, c)\}.$$

Let $C = \{1, 3\}$ and $D = \{a, c\}$. Then

1. domain $f = \{1, 2, 3, 4\}$, codomain $f = \{a, b, c\}$, range $f = \{a, c\}$,
2. $f(C) = \{a\}$, $f^{-1}(D) = \{1, 2, 4\}$,
3. the contour of c is $f^{-1}(c) = \{2, 4\}$, and $f^{-1}(b) = \varnothing$.

Some special types of functions are defined below. We suppose a function $f : A \rightarrow B$:

Everywhere defined: Each of the element in the domain is related to an element in the codomain. That is, for all $a \in A$, $f(a) = b$ for some $b \in B$.

Injective: Each element in the codomain has at most one pre-image. That is, for all $a, c \in A$, if $f(a) = f(c)$, then $a = c$. Injective functions are often called **one-to-one** functions.

Surjective: For each $b \in B$, there exists $a \in A$ such that $f(a) = b$. This implies that the range of f is equal to the codomain. Surjective functions are often called **onto** functions.

Bijective: Everywhere defined, injective and surjective. Bijective functions are also called **one-to-one correspondences** or **bijections**.

Example 2.26 The function defined in Example 2.25 is not everywhere defined since the element 3 has no image. It is not injective because 2 and 4 have the same image. Also, it is not surjective since b has no pre-image.

Example 2.27 The function $f : \mathbb{R} \to \mathbb{R}$ defined by $f(x) = 1/x$ is not everywhere defined because $f(0)$ is undefined. We can, however, let $A = \mathbb{R} \setminus \{0\}$. Then $f : A \to \mathbb{R}$ is everywhere defined. The new definition is also a one-to-one function. Is it onto?

Example 2.28 The rotation function $f : \mathbb{R}^2 \to \mathbb{R}^2$ is given by

$$f(x_1, x_2) = (x_1 \cos\theta - x_2 \sin\theta, x_1 \sin\theta + x_2 \cos\theta),$$

where $\theta \in [0, 2\pi)$. It rotates a point in the anticlockwise direction by the angle θ. It can be shown that the function is a bijection.

Suppose that $f : A \to B$ is a bijection. Since f is one-to-one, the pre-image $f^{-1}(b)$ for every $b \in B$ has at most one element in A. Therefore $f^{-1} : B \to A$ is a function from B to A. Since f is onto, f^{-1} is everywhere defined. Also, f is everywhere defined means that f^{-1} is onto. Finally, since f is a function, f^{-1} is one-to-one. We conclude that f^{-1} is a bijection from B to A. We call f^{-1} the **inverse function** of f. For every $a \in A$ and every $b \in B$,

$$f^{-1}(f(a)) = a, \qquad f(f^{-1}(b)) = b.$$

Let $f : A \to B$ and $g : B \to C$ be functions. Notice that B is the codomain of f and the domain of g. We can define a **composite function** from A to C and denoted by $g \circ f : A \to C$ as

$$(g \circ f)(a) = g(f(a)),$$

for any $a \in A$.

When the domain and the codomain of a function are the same set, we sometimes call the function an **operator** on a set. Suppose that f is an operator on A. An element $p \in A$ is called a **fixed point** of f if $f(p) = p$. Fixed points are important in economic analysis because they represent the equilibria or the steady states of dynamical systems.

Example 2.29 Let $A = \mathbb{R} \setminus \{0\}$ and let $f : A \to A$ be defined as $f(x) = 1/x$. Then f is a bijection. The inverse function is $f^{-1}(y) = 1/y$. The points $p = -1, 1$ are fixed points of f and f^{-1}.

Now let $g : A \to \mathbb{R}$ be defined as $g(x) = x^2$. Then $g \circ f : A \to \mathbb{R}$ becomes

$$(g \circ f)(x) = (1/x)^2 = 1/x^2.$$

Example 2.30 Let $A = \mathbb{R} \setminus \{1\}$ and $B = \mathbb{R} \setminus \{2\}$. Show that $f : A \to B$ defined by

$$f(x) = \frac{2x + 3}{x - 1}$$

is a bijection.

Proof By inspection f is well defined in $\mathbb{R}$ except for the point $x = 1$. Therefore it is everywhere defined. Let $a, b \in A$ and

$$\frac{2a + 3}{a - 1} = \frac{2b + 3}{b - 1}.$$

Then $(2a + 3)(b - 1) = (2b + 3)(a - 1)$, which can be reduced to $a = b$. Therefore f is one-to-one. To show that f is onto, let $y \in B$ and consider $x = (y + 3)/(y - 2)$. Then

$$f(x) = \frac{2(y + 3)/(y - 2) + 3}{(y + 3)/(y - 2) - 1} = \frac{2(y + 3) + 3y - 6}{y + 3 - y + 2} = y.$$

Notice that $f^{-1}(y) = (y + 3)/(y - 2)$. □

Example 2.31 The following are some special functions that we frequently encounter. Here we let $f : A \to B$ be a function.

1. Let $c \in B$. If for all $x \in A$, $f(x) = c$, then we call f a constant function.
2. Let $A = B$. If for all $x \in A$, $f(x) = x$, then f is the identity function.
3. Let $B = \{0, 1\}$ and $S \subseteq A$, a characteristic function of S is defined as

$$f(x) = \begin{cases} 1 & \text{if } x \in S, \\ 0 & \text{if } x \notin S. \end{cases}$$

In econometric analysis, $f(x)$ is often called a dummy variable.
4. The function f is called additive if for all $x, y \in A$,

$$f(x + y) = f(x) + f(y).$$

5. Suppose that the scalar multiplication αx is defined for all $\alpha > 0$ and $x \in A$. Then f is called linearly homogenous if

$$f(\alpha x) = \alpha f(x).$$

An important class of functions called **functionals** or **real-valued functions** is defined as the case that the codomain is $\mathbb{R}$. Since $\mathbb{R}$ is a linear order, it possesses the trichotomy property that for any $x, y \in \mathbb{R}$, either $x < y$, $x = y$ or $x > y$. As a result, it induces an ordering on the domain.

In particular, let $f : S \to \mathbb{R}$. For any $a, b \in S$, we can define an ordering $\succsim$ on S such that

$$(a \succsim b) \Leftrightarrow (f(a) \geq f(b)).$$

Contour sets can be defined using numbers instead of an element in S. For example, the upper contour set

$$\succsim_f (x) = \{a \in S : f(a) \geq x\}$$

consists of all the elements in S that have values greater than or equal to the number x. The subscript f means that the ordering $\succsim$ is induced from f. Similarly the lower contour set of x is

$$\precsim_f (x) = \{a \in S : f(a) \leq x\}.$$

Example 2.32 One of the most important application of functionals in economics is the utility function. A consumer's preference relation on the consumption set $\mathbb{R}_+^n$ is represented by a real-valued utility function $U : \mathbb{R}_+^n \to \mathbb{R}$. The function U is assumed to be everywhere defined so that the preference relation is complete. For consumption bundles $a, b \in \mathbb{R}_+^n$, we say that a is at least as good as b, or $a \succsim b$ if $U(a) \geq U(b)$. Construction of the utility function from a well-defined preference structure can be found in most textbooks in microeconomics. See, for example, Jehle and Reny (2011, 13–16).

Sometimes we want to consider the subset of the product set $S \times \mathbb{R}$ which is above or below the graph of a functional. The geometry of the set can be used to characterize the function. The **epigraph** of a functional $f : S \to \mathbb{R}$ is the subset above the graph and is defined as

$$\text{epi } f = \{(a, x) \in S \times \mathbb{R} : x \geq f(a)\}.$$

Figure 2.2 depicts the epigraph of the function f on the interval (x_1, x_2). On the other hand, the **hypograph** is the subset below the graph:

$$\text{hypo } f = \{(a, x) \in S \times \mathbb{R} : x \leq f(a)\}.$$

It is clear that $(\text{epi } f) \cap (\text{hypo } f) = \text{graph } f$.

A more general idea of order-preserving functions is the class of **monotone functions**. Let $(A, \succsim_A)$ and $(B, \succsim_B)$ be ordered sets. A function $f : A \to B$ is called an **increasing function** if for all $a, b \in A$,

Fig. 2.2 Epigraph of a functional

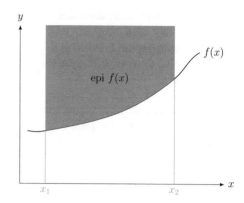

$$(a \gtrsim_A b) \Rightarrow (f(a) \gtrsim_B f(b)).$$

On the other hand, f is called a **decreasing function** if

$$(a \gtrsim_A b) \Rightarrow (f(b) \gtrsim_B f(a)).$$

2.6 Exercises

1. Define the following sets with mathematical symbols:
   (a) The set of odd numbers.
   (b) The set of even numbers.
   (c) The set of all positive multiples of seven.
2. Determine which of the following are sets. Justify your answers.
   (a) $\{x : x$ is a graduate student at your university.$\}$
   (b) $\{p : p$ is a logical statement.$\}$
   (c) $\{x \in \mathbb{R} : x^2 + 1 = 0.\}$
   (d) $\{x : x$ is a right glove.$\}$
   (e) $\{x : x$ is a left chopstick.$\}$
3. Let M denote Mark Messier, E denote the Edmonton Oiler hockey team in 1988 and H denote the National Hockey League. What are the relationships between E, H and M?
4. Let the sets A and B be defined as

$$A = \{x \in \mathbb{R} : x \geq 0\},$$
$$B = \{x = y^2 : y \in \mathbb{R}\}.$$

Show that $A = B$.

5. Prove the following statements:
   (a) For any set S, $\varnothing \subseteq S$.
   (b) The empty set $\varnothing$ is unique.
6. Determine the cardinalities of the following sets.
   (a) $A = \varnothing$
   (b) $B = \{\varnothing\}$
   (c) $C = \{\varnothing, \{\varnothing\}\}$
   (d) $D = \{\varnothing, \{\varnothing\}, \{\varnothing, \{\varnothing\}\}\}$
7. Let

$$A = \{a, b, c, d, e\},$$

$$B = \{1, 2, 3, 4, 5\},$$

$$C = \{x : x \text{ is an alphabet in the word 'concerto'}\},$$

$$P = \{x \in \mathbb{N} : x \text{ is a prime number}\}.$$

Find the sets $A \cup B$, $A \cap C$, $B \cap P$, $A \cap B$, $C \cup B$, $A \setminus C$, $C \setminus A$, $B \setminus P$.
8. Let A and B be subsets of a universal set U. Suppose that $A^c \cup B = U$. Prove that $A \subseteq B$.
9. Let A, B and C be subsets of a universal set U. Show that
   (a) $A \setminus (B \cup C) \subseteq (A \setminus B) \cap (A \setminus C)$,
   (b) $A \Delta B = (A \cup B) \setminus (A \cap B)$,
   (c) $A \cup (U \setminus A) = U$.
10. Let A, B and C be subsets of a universal set U. Show that
    (a) $(A \cap B) \subseteq (A \cup B)$,
    (b) $A \setminus B = A \cap B^c$,
    (c) $A \cup (B \cap C) = (A \cup B) \cap (A \cup C)$.
11. Prove the second De Morgan's law of sets: $(A \cap B)^c = A^c \cup B^c$.
12. Let $A_n = (-1/(1 + n), n]$, $n \in \mathbb{N}$. Find $\bigcap_{n \in \mathbb{N}} A_n$ and $\bigcup_{n \in \mathbb{N}} A_n$.
13. Let $A = \{1, 2\}$. Find A^2. What is $|\mathcal{P}(A^2)|$?
14. Let A, B, C, D be sets. Suppose that $A \subseteq C$ and $B \subseteq D$. Show that $A \times B \subseteq C \times D$.
15. Let A and B be sets. Prove that $A \times B = B \times A$ if and only if $A = B$, or at least one of A or B is empty.
16. Let $(a, b) \in \mathbb{R}_+^2$. Define $A = \{x \in \mathbb{R} : x \geq a\}$ and $B = \{y \in \mathbb{R} : y \geq b\}$. Identify $A \times B$ on a diagram. (In consumer theory, $A \times B$ is called the "no-worse-than set of the bundle (a, b)".)
17. Let $(a, b) \in \mathbb{R}_+^2$. Define $C = \{x \in \mathbb{R}_+ : x \leq a\}$ and $D = \{y \in \mathbb{R}_+ : y \leq b\}$. Identify $C \times D$ on a diagram. (In consumer theory, $C \times D$ is called the "no-better-than set of the bundle (a, b)".)
18. Determine if each of the following relation R on a set S is complete, reflexive, transitive, circular, symmetric, asymmetric or antisymmetric. Justify your answers.
    (a) $S = \{a, b, c\}$, $R = \{(a, a), (b, b), (c, c), (c, a), (b, c), (b, a)\}$.

(b) $S = \{$rock, scissors, paper$\}$, $a\,R\,b$ means "a beats b".

(c) S is a set of people, $a\,R\,b$ means "a is connected to b on Facebook".

(d) S is the set of all commercial and investment banks in an economy, $a\,R\,b$ means "Bank a has an interbank loan from Bank b".

19. Let S be the set of all straight lines in the plane. Define the relation R on S as "is perpendicular to". Determine if the relation is

   (a) complete,
   (b) reflexive,
   (c) transitive,
   (d) circular,
   (e) symmetric,
   (f) equivalence.

   Justify your answers.

20. Let S be the set of all statements. Define a relation "$\Rightarrow$" on S which means "implies". Explain whether the relation is

   (a) complete,
   (b) reflexive,
   (c) symmetric,
   (d) transitive,
   (e) anti-symmetric,
   (f) asymmetric.

21. Let S be a set. Show that the relation $\subseteq$ on the power set of S, $\mathcal{P}(S)$, is a partial order.

22. Determine if the relation R on $\mathbb{N}$ defined as "is a prime factor of" a partial order or not.

23. Suppose that a relation R on a set A is reflexive and circular. Show that it is an equivalence relation.

24. Suppose that a relation on $\mathbb{R}$ is defined by

$$R = \{(x, y) \in \mathbb{R} \times \mathbb{R} : x - y \in \mathbb{Z}\}.$$

That is, $x - y$ is an integer. Prove that R is an equivalence relation.

25. Define the relation

$$R = \{(x, y) \in \mathbb{R}^2 : |x - y| \leq 1\}.$$

Determine if R is an equivalence relation.

26. Define the relation

$$R = \{(m, n) \in \mathbb{Z}^2 : 3|(m - n)\}.$$

Determine if R is an equivalence relation.

27. Suppose that $(S, \sim)$ is an equivalence relation. Prove that for all $a, b \in S, a \sim b$ if and only if $\sim(a) = \sim(b)$.

28. Find the equivalence relation corresponding to the partition defined in Example 2.5.
29. Give the formal definitions of a minimal element and the least element of an ordering $(S, \succsim)$.
30. Consider the ordered set $(\mathbb{R}, \geq)$. Let the set $A = \{-1\} \cup (0, 1)$.
    (a) Find the set of upper bounds of A.
    (b) Find the set of lower bounds of A.
    (c) Find sup A and inf A.
    (d) Does A have a minimal element and a least element?
31. Let $(S, \succsim)$ be an ordering. For all x and y in S, define the following induced relations as
    (a) $x \sim y \Leftrightarrow x \succsim y$ and $y \succsim x$,
    (b) $x \succ y \Leftrightarrow x \succsim y$ and $x \nsim y$.
    Show that if $x \sim y$ and $y \succ z$, then $x \succ z$.
32. Let $(S, \succsim)$ be a complete ordering. Prove that for every $y \in S$,
    (a) $\succsim(y) \cup \prec(y) = S$,
    (b) $\succsim(y) \cap \prec(y) = \varnothing$.
33. Let the set $S = \{1, 2, 3, 4, 5, 6\}$. Define an ordering $(S, \succsim)$ as

$$5 \succsim 3 \succ 1 \succsim 4 \sim 2 \succsim 6.$$

Find the following:
    (a) the interval $[4, 3)$,
    (b) the upper contour set $\succsim(1)$,
    (c) the greatest element of S,
    (d) inf S.
34. Let $(S, \succsim)$ be an ordering. Prove that for every $x, y \in S$, if $x \succsim y$, then $\precsim(y) \subseteq \precsim(x)$.
35. Let $(S, \geq)$ be a linear order and suppose that A is a nonempty bounded subset of S. Show that inf A is unique.
36. Suppose that $\succsim$ is a preference relation on the consumption set $\mathbb{R}^n_+$. Let $a \in \mathbb{R}^n_+$. Is the indifference set $\sim(a)$ a bounded set? If yes find sup $\sim(a)$ and inf $\sim(a)$.
37. Show that for all $n \in \mathbb{N}$, $3|(4^n - 1)$.
38. Show that for all $n \geq 4$, $n^2 \leq n!$. (Recall that n factorial is defined as $n! = n(n-1) \cdots (2)(1)$.)
39. Define the function $f : \mathbb{R} \to \mathbb{R}$ by $y = x^3$.
    (a) Is f everywhere defined, one-to-one and onto? Explain.
    (b) Find the set of all fixed points of f.
    (c) Define the hypograph of f.
40. Define a function $f : X \to \mathbb{R}$ as

$$f(x) = \begin{cases} 1 & \text{if } x \in A, \\ 0 & \text{if } x \notin A, \end{cases}$$

where $A \subseteq X$.

(a) Is f one-to-one?

(b) Is f onto?

(c) What is the graph of f?

(d) What is the hypograph of f?

41. Determine whether the following functions $f : \mathbb{R} \to \mathbb{R}$ are everywhere defined, one-to-one and onto.

(a) $f(x) = \sin x$,

(b) $f(x) = \tan x$,

(c) $f(x) = x^2$.

42. Determine whether the following functions on $\mathbb{R}$ are everywhere defined, one-to-one and onto.

(a) $f(x) = e^x$,

(b) $g(x) = \log x$,

(c) $h(x) = (g \circ f)(x)$.

43. Let $A = \mathbb{R} \setminus \{2\}$ and $B = \mathbb{R} \setminus \{5\}$. Show that $f : A \to B$ defined by

$$f(x) = \frac{5x + 1}{x - 2}$$

is a bijection.

44. Let $f : \mathbb{R} \to \mathbb{R}$ and $g : \mathbb{R} \to \mathbb{R}$ be given by $f(x) = \cos x$ and $g(x) = x^2 + 1$. Find the following:

(a) $g([-1, 1])$;

(b) $(g \circ f)([0, \pi])$;

(c) $(f \circ g)([0, \sqrt{2\pi}])$;

(d) $(g \circ f^{-1})([-\pi, \pi])$.

45. Suppose that f is a function which maps a set X into a set Y.

(a) Show that for any $B \subseteq Y$,

$$f(f^{-1}(B)) \subseteq B.$$

(b) Under what condition that $f(f^{-1}(B)) = B$?

(c) Prove your answer in Part (b).

46. Suppose that A, B and C are sets. Let $\succsim_A$, $\succsim_B$ and $\succsim_C$ be ordered relations on A, B and C respectively.

(a) Define a decreasing function $f : A \to B$.

(b) Suppose that $g : B \to C$ is also a decreasing function. Prove or disprove: the composite function

$$g \circ f : A \to C$$

is increasing.

47. Suppose f and g are increasing functions on $\mathbb{R}$. Show that
    (a) $-f$ is decreasing,
    (b) $f + g$ is increasing,
    (c) $g \circ f$ is increasing.
48. Let f and g be functionals on an ordered set $(S, \succsim)$. Suppose f is increasing and g is decreasing. Prove that $f - g$ is increasing.
49. Let $U : \mathbb{R}_+^n \to \mathbb{R}$ be a utility function. Let $g : \mathbb{R} \to \mathbb{R}$ be a strictly increasing function. Show that $g \circ U$ is also a utility function representing the preference order $\succsim$.

References

Devlin, K. (1993). *The joy of sets*, Second Edition. New York: Springer Science+Business Media.

Economist. (2008). Easy as 1, 2, 3. December 30 issue.

Ellis, G. (2012). On the philosophy of cosmology. Talk at Granada Meeting, 2011. Available at <http://www.mth.uct.ac.za/~ellis/philcosm_18_04_2012.pdf>.

Gerstein, L. J. (2012). *Introduction to mathematical structures and proofs*, Second Edition. New York: Springer Science+Business Media.

Holt, J. (2008). Numbers guy. *The New Yorker*, March 3 issue.

Jehle, G. A., & Reny, P. J. (2011). *Advanced microeconomic theory*, Third edition. Harlow: Pearson Education Limited.

Matso, J. (2007). Strange but true: infinity comes in different sizes. *Scientific American*, July 19 issue.

Rudin, W. (1976). *Principles of mathematical analysis*, Third edition. New York: McGraw-Hill.

Strogatz, S. (2010). Division and its discontents. *The New York Times*, February 21 issue.

Basic Topology

3.1 Introduction to Metric Space

3.1.1 Topological Space

Let S be a set. A **topology** on S is a collection $\mathscr{T}$ of subsets of S, called open sets, which satisfies the following:

1. $\varnothing$ and S are in $\mathscr{T}$,
2. Any unions of sets in $\mathscr{T}$ are in $\mathscr{T}$,
3. Any finite intersections of sets in $\mathscr{T}$ are in $\mathscr{T}$.

For any set S, the collection $\mathscr{T} = \{\varnothing, S\}$ is a trivial topology. On the other hand, $\mathscr{T} = \mathcal{P}(S)$, the power set of S, is called the discrete topology. In economics we often focus on a specific type of topological space called metric space.

3.1.2 Metric Space

A **metric space** consists of a set X, with elements called points, and a metric or distance function $\rho : X^2 \to \mathbb{R}$, which maps a pair of points in X to a real number. The metric satisfies the following axioms: For all $x, y, z \in X$,

1. $\rho(x, y) \geq 0$, $\rho(x, x) = 0$,
2. $\rho(x, y) = \rho(y, x)$,
3. $\rho(x, y) + \rho(y, z) \geq \rho(x, z)$.

The first axiom requires that the metric or distance between two points must be positive, unless $x = y$. The second axiom is about symmetry. It does not matter if we measure the distance from x to y or from y to x. The last one is called triangular

© Springer Nature Switzerland AG 2019
K. Yu, *Mathematical Economics*, Springer Texts in Business and Economics,
https://doi.org/10.1007/978-3-030-27289-0_3

inequality. It is the generalization of the fact that the sum of two sides of a triangle is greater than or equal to the third side. We often denote a metric space by (X, ρ). Any subset of X with the metric ρ also satisfies the above definition and is therefore a metric space itself.

Example 3.1 The most important example of metric spaces is the **Euclidean space** $\mathbb{R}^n$ with the metric

$$\rho(\mathbf{x}, \mathbf{y}) = \left[(x_1 - y_1)^2 + \cdots + (x_n - y_n)^2 \right]^{1/2}, \tag{3.1}$$

where $\mathbf{x} = (x_1, x_2, \ldots, x_n)$ and $\mathbf{y} = (y_1, y_2, \ldots, y_n)$ are n-tuples in $\mathbb{R}^n$. Notice that for real numbers $(n = 1)$ and complex numbers $(n = 2)$, the Euclidean metric is simply the absolute value of the difference between the two points, that is, $\rho(x, y) = |x - y|$. For $n \geq 3$, the absolute value of an n-tuple $\mathbf{x}$ is called the **norm** of $\mathbf{x}$, defined by

$$\|\mathbf{x}\| = (x_1^2 + x_2^2 + \cdots + x_n^2)^{1/2}.$$

Therefore the metric defined in (3.1) can be written as

$$\rho(\mathbf{x}, \mathbf{y}) = \|\mathbf{x} - \mathbf{y}\|.$$

3.1.3   Definitions in Metric Space

In the following definitions let X be a metric space with a metric ρ. Also, A and B are subsets of X.

Diameter: The diameter of A is $\sup\{\rho(x, y) : x, y \in A\}$.

Bounded Set: $A \subseteq X$ is bounded if its diameter is finite.

Open Ball: An open ball of a point p is a set $B_r(p) = \{x \in X : \rho(x, p) < r\}$. The positive number r is called the *radius* of the ball. Sometimes an open ball is called a *neighbourhood* of p.

Limit Point: A point p is called a limit point of A if every open ball $B_r(p)$ contains a point $q \in A$ but $q \neq p$.

Interior Point: A point p is an interior point of A if there exists an open ball of p such that $B_r(p) \subseteq A$.

Boundary Point: A point p is a boundary point of A if every open ball of p contains a point in A and a point not in A. The set of all boundary points of A is called the boundary of A, $b(A)$.

Open Set: $A \subseteq X$ is open if every point of A is an interior point.

Closed Set: $A \subseteq X$ is closed if every limit point of A is a point of A.

Closure: The closure of any set A is $\bar{A} = A \cup b(A)$.

Perfect Set: A closed set A is called perfect if every point in A is a limit point of A.

Dense Set: A is dense in X if for all $x \in X$, x is a limit point of A or $x \in A$.

Separated Sets: A and B are separated if both $A \cap \bar{B}$ and $B \cap \bar{A}$ are empty.

Connected Set: A is said to be connected if it is not a union of two nonempty separated sets.

Compact Set: A compact set in the Euclidean space can be defined as a closed and bounded set.

Convex Set: A subset A in the Euclidean space $\mathbb{R}^n$ is convex if for all $\mathbf{x}, \mathbf{y} \in A$, $\alpha\mathbf{x} + (1 - \alpha)\mathbf{y} \in A$ for $0 < \alpha < 1$.

3.1.4 Some Examples

1. Open sets and closed sets are not exclusive. X and $\varnothing$ are both closed and open.
2. All open balls are open and convex sets.
3. The limit points of a set A do not necessary belong to A. For example, let $A = \{1/n : n = 1, 2, \ldots\}$. The point 0 is not in A but it is a limit point of A. Note that A is neither open nor closed. What is $\bar{A}$?
4. The closure of a set A can also be defined as the union of A and all its limit points.
5. A is closed if and only if $A = \bar{A}$.
6. Separated sets are disjoint, but disjoint sets are not necessary separated. For example, the intervals $[a, b]$ and (b, c) are disjoint but not separated. What about (a, b) and (b, c)?
7. The set of rational numbers $\mathbb{Q}$ is neither a close set nor an open set in $\mathbb{R}$. It is, however, dense in $\mathbb{R}$. Is it connected?
8. Let $X = \mathbb{R}_+^n$. The budget set $B = \{\mathbf{x} : \mathbf{p} \cdot \mathbf{x} \leq M\}$ is compact and convex.[1] What is $b(B)$?
9. A finite set has no limit points and is always compact.
10. An open interval (a, b) is open in $\mathbb{R}$ but not in $\mathbb{R}^2$.
11. The sets $A = [3, \pi]$ and $B = \{-4, e, \pi\}$ are both closed in $\mathbb{R}$. But A is perfect while B is not.

The proofs of some of the above statements can be found in Rudin (1976).

3.1.5 Sequences

A **sequence** is a list of points, $x_1, x_2, \ldots$, in a metric space (X, ρ). If the list is finite, it is similar to an n-tuple. Often we use the notation $\{x_n\}$ to represent an infinite sequence. Formally, an infinite sequence is a function from $\mathbb{N}$ into X, with

---

[1] The dot product $\mathbf{p} \cdot \mathbf{x}$ is defined as $p_1x_1 + p_2x_2 + \cdots + p_nx_n$, which will be discussed in the chapter on linear algebra. The inequality means that total expenditure of a consumer on goods and services cannot exceed income M.

$f(n) = x_n$. Notice that not all points in a sequence need to be distinct. The set of all the points in a sequence is called the **range** of $\{x_n\}$. The sequence is said to be **bounded** if the range has a finite diameter.

A sequence $\{x_n\}$ in a metric space (X, ρ) is said to **converge** to a limit $x \in X$ if for all $\epsilon > 0$, there exists an integer N such that for all $n > N$, $\rho(x_n, x) < \epsilon$. Often a converging sequence is denoted by

$$x_n \to x \quad \text{or} \quad \lim_{n \to \infty} x_n = x.$$

A sequence that does not converge is said to **diverge**.

Example 3.2 The sequence $\{x_n\}$ in $\mathbb{R}$ where $x_n = 1/n$ converges to 0.

Proof Let $\epsilon > 0$ be any positive real number. Let $N = \lceil 1/\epsilon \rceil$. Then for all $n > N$,

$$\rho(x_n, x) = |x_n - 0| = 1/n < 1/N \le \epsilon.$$

Therefore $x_n \to 0$. Notice that the range of the sequence is infinite, and the sequence is bounded.   □

The following result shows that the concept of the limit of a sequence and a limit point of a set are related.

Theorem 3.1 *Suppose that A is a subset of a metric space (X, ρ) and p is a limit point of A. Then there exists a sequence in A that converges to p.*

Proof Since p is a limit point of A, for any $n \in \mathbb{N}$, there is a $x_n \in A$ such that $\rho(x_n, p) < 1/n$. Now for any $\epsilon > 0$, pick N such that $1/N < \epsilon$. Then for all $n \ge N$, we have

$$\rho(x_n, p) < 1/n < 1/N < \epsilon.$$

Therefore $x_n \to p$.   □

A **Cauchy sequence** $\{x_n\}$ in a metric space (X, ρ) is that for all $\epsilon > 0$, there exists an integer N such that $\rho(x_n, x_m) < \epsilon$ if $n \ge N$ and $m \ge N$. It can be shown that any convergent sequence is a Cauchy sequence. If the converse is also true, (X, ρ) is called a **complete metric space**. Euclidean spaces are complete. Complete metric spaces have a number of interesting properties and are very useful in economic analysis. Interested readers can consult Chapter 10 in Royden and Fitzpatrick (2010) for more details.

A sequence $\{x_n\}$ of real numbers is said to be

1. increasing if $x_n \le x_{n+1}$ for all n;
2. decreasing if $x_n \ge x_{n+1}$ for all n.

Increasing and deceasing sequences are called monotone sequences. A useful fact about a monotone sequence is that it converges if and only if it is bounded.

The concept of convergence is important in statistics.

Example 3.3 (Law of Large Number) Let $\bar{X}_n = (X_1 + X_2 + \cdots + X_n)/n$ denote the mean of a random sample of size n from a distribution that has mean μ. Then $\{\bar{X}_n\}$ is a sequence in $\mathbb{R}$. The Law of Large Number states that $\bar{X}_n$ converges to μ in probability, that is, for any $\epsilon > 0$,

$$\lim_{n \to \infty} \Pr(|\bar{X}_n - \mu| < \epsilon) = 1.$$

3.1.6 Series in $\mathbb{R}$

Suppose that $\{a_n\}$ is a sequence in $\mathbb{R}$. The partial sum of $\{a_n\}$ is a sequence $\{s_n\}$ defined by

$$s_n = a_1 + a_2 + \cdots + a_n = \sum_{i=1}^{n} a_i.$$

We also call the sequence $\{s_n\}$ the **infinite series** or simply **series**, which is denoted by $\sum_{n=1}^{\infty} a_n$. We say that the series **converges** to a number $s \in \mathbb{R}$ if

$$\sum_{n=1}^{\infty} a_n = s.$$

Otherwise we say that the series **diverges**. Since $\mathbb{R}$ is a complete metric space, we can apply the Cauchy criterion for convergence on the series. That is, $\sum_{n=1}^{\infty} a_n$ converges if and only if for all $\epsilon > 0$, there exists an integer N such that for $m \geq n > N$,

$$\left| \sum_{i=n}^{m} a_i \right| < \epsilon. \tag{3.2}$$

When $n = m$, the inequality in (3.2) implies that $a_n \to 0$.

Example 3.4 A geometric series is defined as

$$a + ax + ax^2 + ax^3 + \cdots = \sum_{n=0}^{\infty} ax^n,$$

with $a \neq 0$. The sum of the first n terms is

$$s_n = a + ax + ax^2 + \cdots + ax^{n-1}.$$

Multiplying both sides by x, we have

$$xs_n = ax + ax^2 + ax^3 + \cdots + ax^{n-1} + ax^n.$$

Subtracting the above two equations gives

$$s_n - xs_n = a - ax^n,$$

or

$$s_n = \frac{a(1-x^n)}{1-x}.$$

The series converges if $-1 < x < 1$, that is,

$$\sum_{n=0}^{\infty} ax^n = \lim_{n \to \infty} s_n = \lim_{n \to \infty} \frac{a(1-x^n)}{1-x} = \frac{a}{1-x}. \qquad (3.3)$$

An important application in economics is to calculate the present value of an infinite stream of unit incomes. Let $a = 1$ and $x = 1/(1+r)$ where r is the interest rate in each period, (3.3) implies that

$$\sum_{n=0}^{\infty} \frac{1}{(1+r)^n} = \frac{1+r}{r}.$$

You should be able to show that

$$\sum_{n=1}^{\infty} \frac{1}{(1+r)^n} = \frac{1}{r}.$$

Example 3.5 The harmonic series $\sum_{n=1}^{\infty} 1/n$ show that the condition that $a_n \to 0$ is not sufficient for convergence. We have

$$\sum_{n=1}^{\infty} \frac{1}{n} = 1 + \frac{1}{2} + \frac{1}{3} + \frac{1}{4} + \cdots$$

$$= 1 + \frac{1}{2} + \left(\frac{1}{3} + \frac{1}{4} \right) + \left(\frac{1}{5} + \frac{1}{6} + \frac{1}{7} + \frac{1}{8} \right) + \left(\frac{1}{9} + \cdots + \frac{1}{16} \right)$$

$$\geq 1 + \frac{1}{2} + \left(\frac{1}{4} + \frac{1}{4} \right) + \left(\frac{1}{8} + \frac{1}{8} + \frac{1}{8} + \frac{1}{8} \right) + \left(\frac{1}{16} + \cdots + \frac{1}{16} \right)$$

$$= 1 + \frac{1}{2} + \frac{1}{2} + \frac{1}{2} + \cdots .$$

Therefore the series diverges.

Example 3.6 Another interesting series is

$$e = \sum_{n=0}^{\infty} \frac{1}{n!},$$

where $n! = n(n-1)\cdots(2)(1)$ and $0! = 1$. Observe that

$$\sum_{n=0}^{\infty} \frac{1}{n!} = 1 + 1 + \frac{1}{(2)(1)} + \frac{1}{(3)(2)(1)} + \frac{1}{(4)(3)(2)(1)} + \cdots$$

$$< 1 + 1 + \frac{1}{2} + \frac{1}{2^2} + \frac{1}{2^3} + \cdots$$

$$= 3,$$

where we have applied the geometric series in (3.3) to the last equality with $a = 1$ and $x = 1/2$. Therefore the series converges to a number that we call e. It can be shown that e is irrational. An alternative definition is

$$e = \lim_{n \to \infty} \left(1 + \frac{1}{n}\right)^n .$$

Example 3.7 For every real number x, we define the exponential function as[2]

$$e^x = \sum_{n=0}^{\infty} \frac{x^n}{n!} .$$

It can be shown that the series converges for every real number so that the function is everywhere defined in $\mathbb{R}$. The function is increasing so it is one-to-one. Some of the important properties of the exponential function are listed below. For all $x, y \in \mathbb{R}$,

1. $e^x > 0, e^0 = 1$;
2. $e^x \to \infty$ as $x \to \infty$, $e^x \to 0$ as $x \to -\infty$ (see Fig. 3.1);
3. $e^{x+y} = e^x e^y$;
4. the function is differentiable in $\mathbb{R}$ and $d(e^x)/dx = e^x$;

---

[2]The exponential function is also defined for complex numbers. But we shall not pursue the analysis here.

Fig. 3.1 The exponential
function $y = e^x$

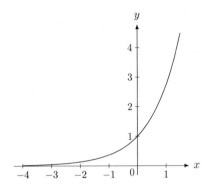

5. the function increases faster than any power function of x, that is,

$$\lim_{x \to \infty} \frac{x^n}{e^x} = 0,$$

for all $n \in \mathbb{N}$;
6. the function is a bijection from $\mathbb{R}$ to $\mathbb{R}_{++}$, the inverse function is the natural log
   function, $f(y) = \log y$.

Example 3.8 Most mathematicians agree that the series

$$\sum_{n=1}^{\infty} n = 1 + 2 + 3 + \cdots$$

diverges to infinity. By exploiting the paradoxes of infinity, some scientists claim
that the series converges to $-1/12$. Overbye (2014) provides an entertaining
description of their claim.

3.2   Continuous Functions

Suppose f is a function from a metric space X to a metric space Y. Let $E \subseteq X$ and
$p \in E$. Then f is **continuous** at p if for every $\epsilon > 0$ there exists a $\delta > 0$ such that

$$\rho_X(x, p) < \delta$$

implies that

$$\rho_Y(f(x), f(p)) < \epsilon$$

for all $x \in E$. If f is continuous at every point in E, then f is said to be continuous on E. A continuous bijection of two metric spaces is called a **homeomorphism**.

Example 3.9 We show that the function $f : \mathbb{R} \to \mathbb{R}$ given by $f(x) = e^x$ with the Euclidean metric is continuous. That is, we have to show that for any $p \in \mathbb{R}$ and $\epsilon > 0$, we can find a $\delta > 0$ such that whenever

$$\rho_X(x, p) = |x - p| < \delta,$$

we have

$$\rho_Y(f(x), f(p)) = |e^x - e^p| < \epsilon.$$

To this end, notice that e^x is an increasing function. If x is between $\log(e^p - \epsilon)$ and $\log(e^p + \epsilon)$, then e^x is between $e^p - \epsilon$ and $e^p + \epsilon$. Therefore we choose

$$\delta = \min\{p - \log(e^p - \epsilon), \log(e^p + \epsilon) - p\}.$$

Then $|e^x - e^p| < \epsilon$ as required.

3.2.1 Properties of Continuous Functions

There are several useful characterizations of a continuous function:

Theorem 3.2 *Let f be a function which maps a metric space X into a metric space Y. The following statements are equivalent:*

1. *The function f is continuous on X.*
2. *For every open set $V \subseteq Y$, the pre-image $f^{-1}(V)$ is open in X.*
3. *For every closed set $W \subseteq Y$, the pre-image $f^{-1}(W)$ is closed.*
4. *Let $\{x_n\}$ be any sequence in X which converges to a point x. Then the sequence $\{f(x_n)\}$ converges to $f(x)$ in Y.*

Proof $(1 \Leftrightarrow 2)$ Suppose f is continuous on X and $V \subseteq Y$ is open. Our goal is to show that every point in $f^{-1}(V)$ is an interior point. So suppose $p \in f^{-1}(V)$ so that $f(p) \in V$. Since V is open, there exists an open ball $B_\epsilon(f(p)) \subseteq V$. In other words, for every point $y \in Y$ such that $\rho_Y(f(p), y) < \epsilon$, $y \in V$. On the other hand, since f is continuous at p, there exist a $\delta > 0$ such that $\rho_Y(f(p), f(x)) < \epsilon$ whenever $\rho_X(p, x) < \delta$. This means that $B_\delta(p) \subseteq f^{-1}(V)$. Hence p is an interior point of $f^{-1}(V)$.

Conversely, suppose $f^{-1}(V) \subseteq X$ is open for every open set $V \subseteq Y$. Let $p \in X$ and $\epsilon > 0$. Let $V = B_\epsilon(f(p))$. Since an open ball is open, $f^{-1}(V)$ is also open. This implies that p is an interior point of $f^{-1}(V)$. By definition there exists $\delta > 0$

such that $\rho_X(p, x) < \delta$ implies that $x \in f^{-1}(V)$. It follows that $f(x) \in V$, or $\rho_Y(f(x), f(p)) < \epsilon$. This shows that f is continuous at p.

$(2 \Leftrightarrow 3)$ W is closed if and only if W^c is open so that $f^{-1}(W^c)$ is open by (2). The proof of statement (4) is left as an exercise.   □

Continuity preserves compactness and connectedness:

Theorem 3.3 *Let f be a continuous function from a metric space X to a metric space Y.*

1. *If $E \subseteq X$ is compact, then $f(E)$ is also compact.*
2. *If $E \subseteq X$ is connected, then $f(E)$ is connected.*

Proof Proof of part 1 can be found in Rudin (1976, p. 89). Here we prove part 2 by contradiction. Assume that E is connected but $f(E)$ is separated, that is, $f(E) = A \cup B$, where $A, B \subset Y$ are nonempty and separated. Let $G = E \cap f^{-1}(A)$ and $H = E \cap f^{-1}(B)$. Notice that $E = G \cup H$ and both G and H are nonempty sets.

Since $G \subseteq f^{-1}(A)$, $G \subseteq f^{-1}(\bar{A})$. By the above corollary, $\bar{A}$ is closed implies that $f^{-1}(\bar{A})$ is also closed. Therefore $\bar{G} \subseteq f^{-1}(\bar{A})$, or $f(\bar{G}) \subseteq \bar{A}$. Since $f(H) = B$ and $\bar{A} \cap B = \varnothing$, we have $f(\bar{G}) \cap f(H) = \varnothing$. This implies that $\bar{G} \cap H = \varnothing$, otherwise any $x \in \bar{G} \cap H$ will have $f(x) \in f(\bar{G}) \cap f(H)$.

By a similar argument we can show that $G \cap \bar{H} = \varnothing$. Together these mean that G and H are separated, which is a contradiction since $E = G \cup H$ and E is connected.   □

If f is a functional, then the compactness property guarantees that f has a maximum value and a minimum value. The following result is sometimes called Weierstrass Theorem.

Theorem 3.4 *Let f be a continuous functional on a metric space X. If $E \subseteq X$ is compact, then there exist p and q in E such that for all $x \in E$,*

$$f(q) \leq f(x) \leq f(p).$$

Proof Let $M = \sup f(E)$ and $m = \inf f(E)$. Since E is compact, then by Theorem 3.3 $f(E)$ is compact. It follows that M and m are in $f(E)$ and the result follows.   □

The connectedness property leads to the important intermediate-value property of continuous functionals.

Theorem 3.5 (Intermediate Value Theorem) *Let f be a continuous functional on the interval $[a, b]$. Suppose that $f(a) < f(b)$. For every number c in the interval $(f(a), f(b))$, there exists a number $x \in (a, b)$ such that $f(x) = c$.*

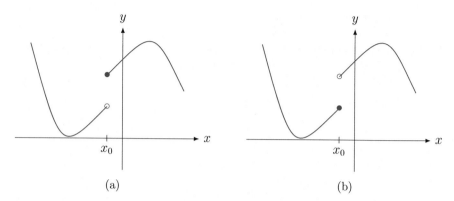

Fig. 3.2 Semicontinuous functions. (**a**) An upper semicontinuous function. (**b**) A lower semicontinuous function

You should be able to prove the following results from the definition of continuity.

Theorem 3.6 *Let f and g be continuous functionals on a metric space X and $\alpha \in \mathbb{R}$. Then $\alpha f, f + g, fg, g \circ f$ and f/g, provided that $g(x) \neq 0$ for all $x \in X$, are continuous on X.*

3.2.2 Semicontinuity

Theorem 3.4 implies that solutions for optimization problems on a compact set always exist if the objective function is continuous. In a typical economic application, however, we only need to find either the maximum or the minimum but not both. In these situations we can relax the continuity requirement a little bit. For example, for a maximum solution to exist, we require the objective function to be upper semicontinuous.

Formally, a functional f on a metric space X is **upper semicontinuous** if for every $\alpha \in \mathbb{R}$, the upper contour set

$$\{x \in X : f(x) \geq \alpha\}$$

is closed. Similarly, f is **lower semicontinuous** if the lower contour set $\{x \in X : f(x) \leq \alpha\}$ is closed for all $\alpha \in \mathbb{R}$. In Fig. 3.2a, for any value of α between the jump at the point x_0, the upper contour set defined above is closed but the lower contour set is not. Therefore the function is upper semicontinuous but not lower semicontinuous. You should convince yourself that the function in Fig. 3.2b is lower semicontinuous. It is straightforward to show that a function is continuous if and only if it is both upper and lower semicontinuous.

3.2.3 Uniform Continuity

Notice that in the definition of continuity above, δ depends on ϵ and the point p. The following definition puts a stricter condition on f:

A function $f : X \to Y$ is **uniformly continuous** on $E \subseteq X$ if for all $\epsilon > 0$ there exists a $\delta > 0$ such that if $p, q \in E$ and

$$\rho_X(p, q) < \delta,$$

then

$$\rho_Y(f(p), f(q)) < \epsilon.$$

Thus δ depends only on ϵ and not on p or q. It is obvious that a uniformly continuous function is continuous. We state the following useful result without proof (see Rosenlicht, 1968, 80–81):

Theorem 3.7 *Let f be a continuous function from a metric space X to a metric space Y. If $E \subseteq X$ is compact, then f is uniformly continuous on E.*

Theorem 3.2 guarantees that the image of a convergent sequence is also convergent under a continuous mapping. The following example shows that this is not necessary true for a Cauchy sequence.

Example 3.10 Define the continuous function $f(x) = \log x$ on $\mathbb{R}_{++} = (0, \infty)$. Consider the Cauchy sequence $\{x_n\} = \{e^{-n}\}$. The image of the sequence with f is $\{y_n\} = \{-n\}$, which is not a Cauchy sequence.

The following two results provide the remedies.

Theorem 3.8 *Let f be a uniformly continuous function which maps a metric space X into a metric space Y and let $\{x_n\}$ be a Cauchy sequence in X. Then the image $\{f(x_n)\}$ is also a Cauchy sequence in Y.*

Theorem 3.9 *Let X be a complete metric space. Then $E \subseteq X$ is also a complete metric space if and only if it is closed.*

In Example 3.10, f is not uniformly continuous on $\mathbb{R}_{++}$. On the other hand, $\mathbb{R}_{++}$ is not complete and so the Cauchy sequence $\{x_n\} = \{e^{-n}\}$ is not convergent.

3.2.4 Fixed Point Theorems

Let X be a metric space and $f : X \to X$. Then $x \in X$ is called a fixed point if $f(x) = x$. In economics a fixed point is an equilibrium point in a dynamical system. Therefore the conditions of existence of fixed point are important. We illustrated the problem with a simple example.

Suppose that $f : [a, b] \to [a, b]$ is continuous. Then there exists at least one fixed point. To show this, let $g(x) = f(x) - x$ on $[a, b]$. Then $g(a) \geq 0 \geq g(b)$. By Theorem 3.5 there exists a point $x \in [a, b]$ such that $g(x) = 0$, which implies that $f(x) = x$. An extension of this result to a more general setting is as follows. Interested readers can consult McLennan (2014) for extensive discussions on fixed point theorems.

Theorem 3.10 (Brouwer Fixed Point Theorem) *Let E be an nonempty, compact and convex subset of an Euclidean space. Then every continuous function $f : E \to E$ has at least one fixed point.*

3.3 Sequences and Series of Functions

In this section we briefly discuss sequences and series of functions, which sometimes appear in probability theory, approximation of functions and dynamical systems.

3.3.1 Pointwise Convergence

Let $\{f_n\}$ be a sequence of functionals defined on a set S. Suppose that for every point $x \in S$, the sequence of numbers $\{f_n(x)\}$ converges. Then we can define a **limit function** $f(x)$ on S such that

$$f(x) = \lim_{n \to \infty} f_n(x).$$

We say that the sequence of function $\{f_n\}$ **converges pointwise** on S. Similarly, if the series $\sum f_n(x)$ converges for every $x \in S$, then we call the function

$$f(x) = \sum_{n=1}^{\infty} f_n(x)$$

the **sum of the series** $\sum f_n(x)$.

Naturally, we are interested in whether properties of each of the functions f_n will be inherited by the limit function or the sum. The following simple example shows that it is not always the case.

Example 3.11 Suppose that $f_n : [0, 1] \to \mathbb{R}$ is defined by

$$f_n(x) = x^n.$$

Then for $0 \leq x < 1$, $f_n(x) \to 0$. When $x = 1$, $f_n(x) \to 1$. Therefore the limit function is

$$f(x) = \begin{cases} 0 & \text{if } 0 \leq x < 1, \\ 1 & \text{if } x = 1. \end{cases}$$

Example 3.11 shows that although each of the functions f_n in the sequence is continuous on the interval $[0, 1]$, the limit function $f(x)$ is not. In other words, when $x = 1$,

$$\lim_{t \to x} \lim_{n \to \infty} f_n(t) \neq \lim_{n \to \infty} \lim_{t \to x} f_n(t).$$

It turns out that properties in differentiation and integration are not preserved under pointwise convergence as well. We need a stricter criterion for convergence that can carry over these properties to the limit function.

3.3.2   Uniform Convergence

A sequence of functionals $\{f_n\}$ defined on a set S is said to **uniformly converge** to a limit function f if for every $\epsilon > 0$, there exists an integer N such that for all $n > N$,

$$|f_n(x) - f(x)| < \epsilon \tag{3.4}$$

for all $x \in S$. The key difference between pointwise convergence and uniform convergence is that in the former N depends on ϵ and x, whereas uniform convergence means that N depends only on ϵ. In other words, inequality (3.4) is satisfied uniformly for all $x \in S$ once N is chosen. A formal definition for uniform convergence of the series $\sum f_n(x)$ is left as an exercise. Not surprisingly, the Cauchy criterion applies to uniform convergence.

Theorem 3.11 *A sequence of functionals $\{f_n\}$ defined on a set S uniformly converges to a limit function f on S if and only if for every $\epsilon > 0$, there exists an integer N such that for all $m, n > N$,*

$$|f_n(x) - f_m(x)| < \epsilon$$

for all $x \in S$.

Example 3.12 Show that the sequence of functions defined by

$$f_n(x) = \frac{x}{1 + nx^2}$$

converges uniformly to a function f for every $x \in \mathbb{R}$.

Proof Observe that for every $x \in \mathbb{R}$,

$$\left| \frac{x}{1 + x^2} \right| = \frac{|x|}{1 + x^2} < 1.$$

For any $\epsilon > 0$, let $N > 1/\epsilon$. Then for all $n > N$,

$$|f_n(x)| = \left| \frac{x}{1 + nx^2} \right|$$

$$= \frac{1}{n} \left(\frac{|x|}{1/n + x^2} \right)$$

$$\leq \frac{1}{n} \left(\frac{|x|}{1 + x^2} \right)$$

$$< \frac{1}{n} < \frac{1}{N} < \epsilon.$$

This shows that $\{f_n(x)\}$ converges uniformly to $f(x) = 0$ on $\mathbb{R}$. $\qquad\square$

Example 3.13 (Central Limit Theorem) Let $\bar{X}_n = (X_1 + X_2 + \cdots + X_n)/n$ denote the mean of a random sample of size n from a distribution that has mean μ and variance σ^2. Let

$$Z_n = \frac{\sqrt{n}(\bar{X}_n - \mu)}{\sigma}.$$

Then $\{Z_n\}$ is a sequence of random variables with probability density functions $\{f_n\}$. The Central Limit Theorem states that $\{f_n\}$ converges uniformly to the probability density function of a normal distribution with mean 0 and variance 1. That is, the limit function is

$$f(x) = \frac{1}{\sqrt{2\pi}} e^{-x^2/2}.$$

This is a remarkable result because any distribution with a finite mean and variance converges to the standard normal distribution, and knowledge of the functional form of f_n is not necessary.

Theorem 3.12 *Suppose that* $\{f_n\}$ *is a sequence of continuous functions that converges uniformly to a limit function* f *on a set* S. *Then* f *is continuous on* S.

Interested readers can consult Rudin (1976, chapter 7) for a proof of this result and the relation between uniform convergence and differentiation and integration.

For series of functions, there is a test for uniform convergence.

Theorem 3.13 (Weierstrass M-Test) *Let* $\{f_n\}$ *be a sequence of functions defined on a set* S *and let* $\{M_n\}$ *be a sequence of real numbers. Suppose that for all* $x \in S$ *and all* $n \in \mathbb{N}$,

$$|f_n(x)| \leq M_n.$$

If $\sum M_n$ *converges, then* $\sum f_n(x)$ *converges uniformly on* S.

Proof The proof uses the Cauchy criterion for series of functions. That is, suppose $\sum M_n$ converges. Then for any $\epsilon > 0$, there exists an integer N such that for all $m, n > N$,

$$\left| \sum_{i=m}^{n} f_i(x) \right| \leq \sum_{i=m}^{n} M_i < \epsilon,$$

for all $x \in S$. Therefore $\sum f_n(x)$ converges uniformly on S. $\qquad\square$

Example 3.14 In a basic centralized model in macroeconomics, the economy seeks to maximize a series of utility function $\sum_{t=1}^{\infty} \beta^t U(c_t)$, where U is the instantaneous utility function of consumption c_t, and $\beta \in (0, 1)$ is a discount factor. We use t as the counter instead of n to reflect that it represents time periods. The constraint of the economy in each period is

$$F(k_t) = c_t + k_{t+1} - (1 - \delta)k_t, \tag{3.5}$$

where F is the aggregate production function, k_t is the capital stock and δ is the depreciation rate of capital. The function F is usually assumed to be increasing and concave, reflecting decreasing returns to scale. Then the economy will converge to an equilibrium $k_t = k^*$. From (3.5) the maximum consumption is $c_M = F(k^*)$, which means that the economy consumes the whole output and makes zero investment, that is,

$$k_{t+1} - (1 - \delta)k_t = 0.$$

Define $M_t = \beta^t U(c_M)$. Then the geometric series $\sum M_t$ converges. Since by definition $\beta^t U(c_t) \leq M_t$, for all t, by Theorem 3.13 the series of utility functions $\sum_{t=1}^{\infty} \beta^t U(c_t)$ converges uniformly to a limit function. If the instantaneous utility function U is continuous, then by Theorem 3.12 the limit function is continuous. It follows from Theorem 3.4 that a maximum value exists. The technique of dynamic optimization will be discussed in Chap. 9.

3.4 Exercises

1. Let $S = \{a, b, c\}$. Show that

$$\mathcal{T} = \{\varnothing, \{a, b\}, \{b\}, \{b, c\}, S\}$$

   is a topology on S.
2. Let X be an infinite set. For $x, y \in X$, define

$$\rho(x, y) = \begin{cases} 1 & \text{if } x \neq y, \\ 0 & \text{if } x = y. \end{cases}$$

   Prove that (X, ρ) is a metric space. Give an example of proper subset of X that is open. Are there any closed sets and compact sets?
3. Show that the set $\mathbb{R}^n$ with the distance function $\rho(\mathbf{x}, \mathbf{y}) = \max_{i=1}^{n} |x_i - y_i|$ is a metric space.
4. (Symbolic dynamics) Suppose that the set Ω_2 contains all bi-infinite sequences of two symbols, say 1 and 2. A point $a \in \Omega_2$ is the sequence

$$a = \{\ldots, a_{-2}, a_{-1}, a_0, a_1, a_2, \ldots\},$$

   where $a_i = \{1, 2\}$. The position of a_0 must be specified. Otherwise, a sequence such as

$$a = \{\ldots, 2, 1, 2, 1, 2, 1, \ldots\}$$

   can represent two distinct sequences, one with $a_0 = 2$ and the other with $a_0 = 1$. The metric between two points $a, b \in \Omega_2$ is defined by

$$\rho(a, b) = \sum_{i=-\infty}^{\infty} \frac{\delta(a_i, b_i)}{2^{|i|}},$$

   where

$$\delta(a_i, b_i) = \begin{cases} 0 & \text{if } a_i = b_i, \\ 1 & \text{if } a_i \neq b_i. \end{cases}$$

   Show that (Ω_2, ρ) is a metric space.

5. Determine whether the following sets are (i) open, (ii) closed, (iii) both open and closed, or (iv) neither open nor closed:

   (a) $\mathbb{Z} \subseteq \mathbb{R}$

   (b) $\{(x, y) : 1 < x < 2, y = x\} \subseteq \mathbb{R}^2$,

   (c) $\{x \in \mathbb{Q} : 1 \leq x \leq 2\} \subseteq \mathbb{R}$

6. Prove or disprove: Let A and B be connected in a metric space. Then there exists a boundary point in one of the sets which belongs to the other set.

7. Consider the Euclidean metric on $\mathbb{R}$. Let $A = (-1, 1) \cup (1, 2)$, that is, A is the union of two intervals. Answer the following questions with explanations.

   (a) What is the diameter of A?

   (b) Is A open, closed or neither?

   (c) Is the point $x = 1$ a limit point of A?

   (d) What is $\bar{A}$?

   (e) Is A connected?

8. Consider the Euclidean metric space $\mathbb{R}^2$. Let $A = \{\mathbf{x} \in \mathbb{R}^2 : \rho(\mathbf{x}, \mathbf{0}) \leq 1\}$.

   (a) What is the diameter of A?

   (b) What is $b(A)$?

   (c) What is $\bar{A}$?

   (d) Is A bounded?

   (e) Is A open, closed, neither or both?

   (f) Is A perfect, separated, compact and convex?

9. Let $\mathbb{R}$ be the set of real numbers. Define a metric ρ on $\mathbb{R}$ as

$$\rho(x, y) = \begin{cases} 1 \text{ if } x \neq y, \\ 0 \text{ if } x = y. \end{cases}$$

   Let $A = \{1/n : n = 1, 2, 3, \ldots\}$ and $B = \{-4, e, \pi\}$. Answer the following questions with explanation.

   (a) Is the point 0 a limit point of A?

   (b) Is A an open set, closed set, both or neither?

   (c) Is B an open set, closed set, both or neither?

   (d) What is the diameter of B?

10. Consider the Euclidean space $\mathbb{R}^2$. Let $A = \{(x, 0) : a < x < b\}$, that is, A is an open interval on the horizontal axis. Determine and explain whether A is

    (a) open,

    (b) closed,

    (c) perfect,

    (d) compact.

11. Prove that a set is closed if and only if its complement is open.

12. Show that

    (a) the intersection of two closed sets is closed,

    (b) the union of two closed sets is closed,

    (c) the intersection of two open sets is open,

    (d) the union of two open sets is open.

13. Let (X, ρ) be a metric space. Suppose that A and B are separated subsets of X. Prove or disprove:

$$\bar{A} \subseteq B^c.$$

14. Define the sequence $\{x_n\}$ as $\{1, 1/2, 1/4, 1/6, \ldots\}$.
    (a) What is the range of $\{x_n\}$?
    (b) Is $\{x_n\}$ bounded?
    (c) Let A be the range of $\{x_n\}$. Find sup A.
    (d) Find inf A.
15. Prove or disprove that the following sequences $\{x_n\}$ in $\mathbb{R}$ converge:
    (a) $x_n = 1/\sqrt{n}$.
    (b) $x_n = n^2$.
    (c) $x_n = (-1)^n$.
    (d) $x_n = 1 + (-1)^n/n$.
    In each case determine if the sequence is bounded and if the range is finite.
16. Let $\{x_n\}$ be sequence in a metric space (X, ρ) which converges to a point a. Let A denotes the range of $\{x_n\}$. Is a always a limit point of the set A?
17. Show that the limit of a sequence is unique.
18. Let $\{x_n\}$ and $\{y_n\}$ be two sequences in a metric space (X, ρ).
    (a) Show that $\lim_{n \to \infty} x_n = x$ if and only if for all $r > 0$, the open ball $B_r(x)$ contains all but finitely many terms of $\{x_n\}$.
    (b) Suppose $\lim_{n \to \infty} x_n = x$ and $\rho(x_n, y_n) \to 0$. Show that $\lim_{n \to \infty} y_n = x$.
    (c) Consider the case that $X = \mathbb{R}$ so that the metric is $\rho(x, y) = |x - y|$. Suppose $\lim_{n \to \infty} x_n = x$ and $\lim_{n \to \infty} y_n = y$. Show that $\lim_{n \to \infty}(x_n + y_n) = x + y$.
19. Show that a convergent sequence in a metric space is a Cauchy sequence.
20. Consider the natural order $\geq$ on the set of rational numbers $\mathbb{Q}$. Define a sequence $\{x_n\}$ in $\mathbb{Q}$ as follows:

$$x_1 = 3,$$
$$x_2 = 3.1,$$
$$x_3 = 3.14,$$
$$x_4 = 3.141,$$
$$x_5 = 3.1415,$$
$$x_6 = 3.14159,$$
$$\vdots$$

That is, x_{n+1} includes one more digit from the number π than x_n. Note that $\pi \notin \mathbb{Q}$. Let A be the range of $\{x_n\}$.
    (a) What is the set of upper bounds of A?

(b) Does sup A exists in $\mathbb{Q}$? If yes, what is it?

(c) Does inf A exists in $\mathbb{Q}$? If yes, what is it?

21. Let $\{x_n\}$ be an increasing sequence. Suppose $\{x_n\}$ is bounded. Prove that the sequence converges by following the hints below:

(a) Let E be the range of $\{x_n\}$ and let $x = \sup E$.

(b) Show that for every $\epsilon > 0$, there exists an integer N such that $x-\epsilon < x_N < x$.

(c) Now since $\{x_n\}$ is increasing, $n > N$ implies that $x - \epsilon < x_n < x$.

22. Find

$$\sum_{n=0}^{\infty} \left(\frac{r}{1+r}\right)^n,$$

where r is the market interest rate.

23. Show that the function $f : \mathbb{R} \to \mathbb{R}$ given by $f(x) = x^2$ with the Euclidean metric is continuous. Hint: you may find this useful:

$$|x^2 - p^2| = |(x + p)(x - p)|$$
$$= |(x - p + 2p)(x - p)|$$
$$\leq (|x - p| + 2|p|)|x - p|.$$

24. Show that the rotation function defined in Example 2.28 of Chap. 2 is continuous.

25. Show that the function $f : \mathbb{R} \to \mathbb{R}$ given by

$$f(x) = \begin{cases} 1 & \text{if } x \text{ is rational,} \\ 0 & \text{if } x \text{ is irrational,} \end{cases}$$

is nowhere continuous.

26. Let $\bar{B}_r(p) = \{x \in X : \rho(x, p) \leq r\}$ be a closed ball centred at p with radius r in a metric space X. Let f be a continuous functional on X. Show that $f(\bar{B}_r(p))$ is bounded.

27. Suppose that $f : X \to \mathbb{R}$ is continuous. Prove that $\succsim_f (a) = \{x : f(x) \geq a\}$ is closed.

28. Suppose that S is a compact set in a metric space X. Show that every continuous functional $f : S \to \mathbb{R}$ is bounded, that is, $f(S)$ is bounded.

29. Let $u : X \to \mathbb{R}$ be a utility function on an ordered set $(X, \succsim)$. Let $g : \mathbb{R} \to \mathbb{R}$ be a strictly increasing function. Show that $g \circ u$ is also a utility function representing the order $\succsim$.

30. Let $f : \mathbb{R} \to \mathbb{R}$ be defined by

$$f(x) = \begin{cases} x^2 & \text{if } x \neq 1, \\ 10 & \text{if } x = 1. \end{cases}$$

(a) Is f lower semi-continuous?

(b) Is f upper semi-continuous?

Provide brief explanations to your answers.

31. Suppose that $f : \mathbb{R}^2 \to \mathbb{R}$ is given by

$$f(x, y) = \begin{cases} x^2 + y^2 & \text{if } (x, y) \neq (0, 0), \\ 2 & \text{if } (x, y) = (0, 0). \end{cases}$$

(a) Define the graph of f.

(b) Is f upper semi-continuous? Explain.

(c) Is f lower semi-continuous?

32. Suppose X, Y, Z are metric spaces and $f : X \to Y$ and $g : Y \to Z$ are continuous functions. Prove that $g \circ f : X \to Z$ is continuous.

33. Prove Theorem 3.8.

34. Prove Theorem 3.9.

35. Let $S = [0, 1] \in \mathbb{R}$. Show that a continuous function $f : S \to S$ has at least one fixed point.

36. Give a formal definition of uniform convergence for the series of functions $\sum f_n(x)$ on a set S.

37. Show that the sequence of functions $\{f_n\}$ where

$$f_n(x) = \frac{1}{1 + x^n}$$

converges to a limit function on $S = [0, 1]$. Does it converge uniformly?

38. Show that

$$f_n(x) = \frac{\cos nx}{\sqrt{n}}$$

converges uniformly to a limit function.

References

McLennan, A. (2014). *Advanced fixed point theory for economics*. Available at <http://cupid.economics.uq.edu.au/mclennan/Advanced/advanced_fp.pdf>.

Overbye, D. (2014). In the end, it all adds up to $-1/12$. *The New York Times*, February 3 issue.

Rosenlicht, M. (1968). *Introduction to analysis*. Scott, Foresman and Co. (1986 Dover edition).

Royden, H. L., & Fitzpatrick, P. M. (2010). *Real analysis*. Boston: Prentice Hall.

Rudin, W. (1976). *Principles of mathematical analysis*, Third edition. New York: McGraw-Hill.

Linear Algebra

4

Linear algebra is the starting point of multivariate analysis, due to its analytical and computational simplicity. In many economic applications, a linear model is often adequate. Even in a more realistic nonlinear model, variables are linearized at the point of interest, usually at a steady-state equilibrium, to study their behaviours in the neighbourhood. In this chapter we first study the basic properties of vector spaces. Then we turn our attention to transformations between vector spaces.

4.1 Vector Spaces

In Chap. 3 we study "distance" between objects in a set, the behaviour of sequences of those objects and functions between spaces that preserve proximity. A vector space is an abstract idea of combining objects in a set in a linear fashion. The objects are called **vectors**, and by **linear combination** we mean that we can scale vectors up or down by multiplying them with scalars, usually real or complex numbers, and then combine them to form another vector. In this chapter we take the scalars to be real numbers.

Formally, a **vector space** V is a set of objects called **vectors** with two operations. The first is called **vector addition**, which combines two vectors to form a third one. That is, for any $\mathbf{x}, \mathbf{y} \in V$, we call $\mathbf{x} + \mathbf{y} \in V$ the **sum** of $\mathbf{x}$ and $\mathbf{y}$.

The second operation is **scalar multiplication**, which multiplies a real number to a vector to form another vector. That is, for any $\mathbf{x} \in V$ and $\alpha \in \mathbb{R}$, the **product** $\alpha\mathbf{x}$ is in V.

Vector additions and scalar multiplications in a vector space follow eight rules or axioms. For any $\mathbf{x}, \mathbf{y}, \mathbf{z} \in V$ and $\alpha, \beta \in \mathbb{R}$:

1. Commutative in addition: $\mathbf{x} + \mathbf{y} = \mathbf{y} + \mathbf{x}$.
2. Associative in addition: $(\mathbf{x} + \mathbf{y}) + \mathbf{z} = \mathbf{x} + (\mathbf{y} + \mathbf{z})$.
3. There exists a unique zero vector $\mathbf{0} \in V$ such that $\mathbf{x} + \mathbf{0} = \mathbf{x}$.

© Springer Nature Switzerland AG 2019
K. Yu, *Mathematical Economics*, Springer Texts in Business and Economics,
https://doi.org/10.1007/978-3-030-27289-0_4

4. For any $\mathbf{x} \in V$, there exist a vector $-\mathbf{x} \in V$ such that $\mathbf{x} + (-\mathbf{x}) = \mathbf{0}$.
5. $1\mathbf{x} = \mathbf{x}$.
6. $(\alpha\beta)\mathbf{x} = \alpha(\beta\mathbf{x})$.
7. $\alpha(\mathbf{x} + \mathbf{y}) = \alpha\mathbf{x} + \alpha\mathbf{y}$.
8. $(\alpha + \beta)\mathbf{x} = \alpha\mathbf{x} + \beta\mathbf{x}$.

Axioms 1 and 2 imply that the order of vector addition is not important. For example,

$$\mathbf{x} + \mathbf{y} + \mathbf{z} = \mathbf{z} + \mathbf{x} + \mathbf{y}.$$

In Axiom 8, the addition symbol on the left-hand side is for real numbers, while the one on the right is for vectors.

Example 4.1 Let $V = \mathbb{R}^n$. For any n-tuples $\mathbf{x}, \mathbf{y} \in \mathbb{R}^n$, where

$$\mathbf{x} = (x_1, x_2, \ldots, x_n),$$

$$\mathbf{y} = (y_1, y_2, \ldots, y_n),$$

vector addition is defined by

$$\mathbf{x} + \mathbf{y} = (x_1 + y_1, x_2 + y_2, \ldots, x_n + y_n).$$

For any $\alpha \in \mathbb{R}$, scalar multiplication is defined by

$$\alpha\mathbf{x} = (\alpha x_1, \alpha x_2, \ldots, \alpha x_n).$$

It is straightforward to verify that the definitions satisfy the eight axioms. Figure 4.1 illustrates what we learned in high school about vector addition. Vectors are represented as arrows, which are defined by their length and direction. To find $\mathbf{a} + \mathbf{b}$, we move the tail of vector $\mathbf{b}$ to the head of vector $\mathbf{a}$, while keeping the direction unchanged. The vector from the tail of $\mathbf{a}$ to the head of $\mathbf{b}$ is $\mathbf{a} + \mathbf{b}$. If we instead put the tail of $\mathbf{a}$ on the head of $\mathbf{b}$, we get $\mathbf{b} + \mathbf{a}$. And as the diagram shows, the results confirm Axiom 1, that vector additions are commutative. The scalar multiplication

Fig. 4.1 Vector addition in $\mathbb{R}^2$

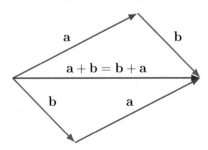

$\alpha\mathbf{a}$, on the other hand, means that the length of $\mathbf{a}$ is multiplied by α. If α is negative, then $\alpha\mathbf{a}$ and $\mathbf{a}$ have opposite directions.

Example 4.2 Let $V = P(n)$ be the set of all polynomials of degree n. For any $f, g \in P(n)$, we have

$$f(x) = a_0 + a_1 x + a_2 x^2 + \cdots + a_n x^n,$$
$$g(x) = b_0 + b_1 x + b_2 x^2 + \cdots + b_n x^n,$$

where $a_i, b_i \in \mathbb{R}$, $i = 0, 1 \ldots, n$. The sum $f + g$ is defined by

$$(f + g)(x) = (a_0 + b_0) + (a_1 + b_1)x + (a_2 + b_2)x^2 + \cdots + (a_n + b_n)x^n.$$

Scalar multiplication is defined by $\alpha f \in P(n)$ such that

$$(\alpha f)(x) = (\alpha a_0) + (\alpha a_1)x + (\alpha a_2)x^2 + \cdots + (\alpha a_n)x^n.$$

In fact, the operations are the same as those in Example 4.1 if we let $(a_0, a_1, \ldots, a_n)$ and $(b_0, b_1, \ldots, b_n)$ be $(n + 1)$-tuples.

Example 4.3 Let $V = S$ be the set of all sequences in $\mathbb{R}$. For all $\{x_n\}, \{y_n\} \in S$ and $\alpha \in \mathbb{R}$, we can define

$$\{x_n\} + \{y_n\} = \{x_n + y_n\},$$

and

$$\alpha\{x_n\} = \{\alpha x_n\}.$$

Then S is a vector space. If we narrow the definition of S to the set of convergent sequences, is it a vector space?

Example 4.4 Define $V = \mathbb{M}^{m \times n}$ as the set of all matrices on real numbers with m rows and n columns. That is, any $A \in \mathbb{M}^{m \times n}$ is in the form

$$A = \begin{pmatrix} a_{11} & a_{12} & \cdots & a_{1n} \\ a_{21} & a_{22} & \cdots & a_{2n} \\ \vdots & \vdots & \ddots & \vdots \\ a_{m1} & a_{m2} & \cdots & a_{mn} \end{pmatrix}.$$

Vector addition is simply matrix addition:

$$A + B = \begin{pmatrix} a_{11} & a_{12} & \cdots & a_{1n} \\ a_{21} & a_{22} & \cdots & a_{2n} \\ \vdots & \vdots & \ddots & \vdots \\ a_{m1} & a_{m2} & \cdots & a_{mn} \end{pmatrix} + \begin{pmatrix} b_{11} & b_{12} & \cdots & b_{1n} \\ b_{21} & b_{22} & \cdots & b_{2n} \\ \vdots & \vdots & \ddots & \vdots \\ b_{m1} & b_{m2} & \cdots & b_{mn} \end{pmatrix}$$

$$= \begin{pmatrix} a_{11} + b_{11} & a_{12} + b_{12} & \cdots & a_{1n} + b_{1n} \\ a_{21} + b_{21} & a_{22} + b_{22} & \cdots & a_{2n} + b_{2n} \\ \vdots & \vdots & \ddots & \vdots \\ a_{m1} + b_{m1} & a_{m2} + b_{m2} & \cdots & a_{mn} + b_{mn} \end{pmatrix}.$$

And scalar multiplication is

$$\alpha A = \begin{pmatrix} \alpha a_{11} & \alpha a_{12} & \cdots & \alpha a_{1n} \\ \alpha a_{21} & \alpha a_{22} & \cdots & \alpha a_{2n} \\ \vdots & \vdots & \ddots & \vdots \\ \alpha a_{m1} & \alpha a_{m2} & \cdots & \alpha a_{mn} \end{pmatrix}.$$

In fact, the operations are analogous to those of mn-tuples.

Example 4.5 Let $V = F(S, \mathbb{R})$ be the set of all functionals on a set S. For all $f, g \in F(S, \mathbb{R})$, vector addition is defined by

$$(f + g)(x) = f(x) + g(x),$$

for all $x \in S$. Scalar multiplication is defined by

$$(\alpha f)(x) = (\alpha) f(x).$$

From the above eight axioms, we can establish a few basic results in vector spaces. Let us label the axioms as A1 to A8. Now, if α is a scalar and $\mathbf{0}$ is the zero vector, then by A3 and A7

$$\alpha \mathbf{0} = \alpha(\mathbf{0} + \mathbf{0}) = \alpha \mathbf{0} + \alpha \mathbf{0}.$$

Adding $-(\alpha \mathbf{0})$ to both sides and using A4, we get

$$\alpha \mathbf{0} = \mathbf{0}. \tag{4.1}$$

That is, any scalar multiplication of the zero vector gives you the zero vector. Similarly, we can show that

$$0\mathbf{x} = \mathbf{0}. \tag{4.2}$$

That is, any vector multiplied by the scalar 0 becomes the zero vector. Observe that

$$0\mathbf{x} = (0 + 0)\mathbf{x} = 0\mathbf{x} + 0\mathbf{x}$$

by (A8). Adding the vector $-0\mathbf{x}$ to both sides gives you

$$-0\mathbf{x} + 0\mathbf{x} = -0\mathbf{x} + 0\mathbf{x} + 0\mathbf{x},$$

which reduces to $\mathbf{0} = \mathbf{0} + 0\mathbf{x}$ by (A4). Applying (A3) to the right-hand side of the equation gives (4.2).

Next we show that if $\alpha\mathbf{x} = \mathbf{0}$, then either $\alpha = 0$ or $\mathbf{x} = \mathbf{0}$. If $\alpha \neq 0$, then by (4.1)

$$\frac{1}{\alpha}(\alpha\mathbf{x}) = \mathbf{0}.$$

But by A5 and A6

$$\frac{1}{\alpha}(\alpha\mathbf{x}) = \left(\frac{1}{\alpha}\alpha\right)\mathbf{x} = 1\mathbf{x} = \mathbf{x}.$$

Therefore we conclude that $\mathbf{x} = \mathbf{0}$.

Another result is that for every vector $\mathbf{x} \in V$,

$$(-\alpha)\mathbf{x} = -(\alpha\mathbf{x}). \tag{4.3}$$

The reason is

$$\mathbf{0} = 0\mathbf{x} \quad \text{(by (4.2))}$$
$$= (\alpha - \alpha)\mathbf{x}$$
$$= \alpha\mathbf{x} + (-\alpha)\mathbf{x} \quad \text{(using A8)}$$

and by A4 the result in (4.3) follows. Consequently, we can write $\mathbf{y} + (-\mathbf{x})$ as $\mathbf{y} - \mathbf{x}$ without ambiguity.

Theorem 4.1 *Suppose $\mathbf{x} \in V$ and $\mathbf{x} \neq \mathbf{0}$. If $\alpha\mathbf{x} = \beta\mathbf{x}$, then $\alpha = \beta$.*

Proof By A4 there exist a vector called $-\alpha\mathbf{x}$. So adding this vector to both sides of $\alpha\mathbf{x} = \beta\mathbf{x}$ gives

$$\alpha\mathbf{x} + (-\alpha\mathbf{x}) = \beta\mathbf{x} + (-\alpha\mathbf{x}). \tag{4.4}$$

It follows that

$$\mathbf{0} = \alpha\mathbf{x} + (-\alpha\mathbf{x}) \quad \text{(by A4)}$$

$$= \beta\mathbf{x} + (-\alpha\mathbf{x}) \quad \text{(from (4.4))}$$
$$= \beta\mathbf{x} + (-\alpha)\mathbf{x} \quad \text{(using (4.3))}$$
$$= (\beta - \alpha)\mathbf{x} \quad \text{(by A8)}$$

Since $\mathbf{x} \neq \mathbf{0}$, by the result above $\beta - \alpha$ must be 0, from which we conclude that $\alpha = \beta$. □

4.2 Basic Properties of Vector Spaces

In this section we characterize some common properties of vector spaces. Suppose that $A = \{\mathbf{x}_1, \mathbf{x}_2, \ldots, \mathbf{x}_n\}$ is a set of vectors in a vector space V. Then $\mathbf{y} \in V$ is a **linear combination** of A if there exists a set of scalars $\{\alpha_1, \alpha_2, \ldots, \alpha_n\}$ such that

$$\mathbf{y} = \alpha_1\mathbf{x}_1 + \alpha_2\mathbf{x}_2 + \cdots + \alpha_n\mathbf{x}_n = \sum_{i=1}^{n} \alpha_i\mathbf{x}_i.$$

The set of all linear combination of A is called the **linear span** or **linear hull** of A. That is, we define

$$\text{span}(A) = \left\{ \sum_{i=1}^{n} \alpha_i\mathbf{x}_i \in V : \alpha_i \in \mathbb{R}, \ i = 1, 2, \ldots, n \right\}.$$

In general, the set A can be countable or uncountable.

Example 4.6 Suppose that $V = \mathbb{R}^3$. Let $A = \{(1, 0, 0), (0, 3, 0)\}$. Then span($A$) is the x-y plane in the three-dimensional space. That is,

$$\text{span}(A) = \{(x, y, 0) \in \mathbb{R}^3 : x, y \in \mathbb{R}\}.$$

This is because for any $(x, y, 0) \in \mathbb{R}^3$, there exists $\alpha, \beta \in \mathbb{R}$ such that

$$(x, y, 0) = \alpha(1, 0, 0) + \beta(0, 3, 0).$$

Example 4.7 Consider the vector space $P(n)$ of all n-degree polynomials in Example 4.2. Let $B = \{f_0, f_1, \ldots, f_n\}$ be the subset of polynomials defined by

$$f_i(x) = x^i, \quad i = 0, 1, \ldots, n.$$

Then span $(B) = P(n)$. In this case we say that the set B **spans** $P(n)$.

Let S be a nonempty subset of a vector space V. We call S a **subspace** of V if S satisfies the definition of a vector space. Recall that any nonempty subset of a metric space is itself a metric space. This, however, does not work in a vector space. This is because any subspace of a vector space has to be "closed" under vector addition and scalar multiplication, as the following theorem shows.

Theorem 4.2 *A set S is a subspace of a vector space V if and only if for any vectors* $\mathbf{x}, \mathbf{y} \in S$ *and any scalar* α, *the linear combination* $\alpha\mathbf{x} + \mathbf{y}$ *is in S.*

Proof Suppose that S is a subspace of V. Since by definition any vector $\mathbf{x}$ and $\mathbf{y}$ in S satisfy the axioms of a vector space, the vector $\alpha\mathbf{x} + \mathbf{y}$ is in S.

Conversely, suppose that S is a nonempty subset of V and, for any vectors $\mathbf{x}, \mathbf{y} \in S$ and any scalar α, $\alpha\mathbf{x} + \mathbf{y}$ is in S. Axioms 1 and 2 are by definition satisfied. For any $\mathbf{x} \in S$, let $\alpha = -1$ so that $(-1)\mathbf{x} + \mathbf{x} = -\mathbf{x} + \mathbf{x} = \mathbf{0}$ is in S. Therefore Axioms 3 and 4 are satisfied. Axioms 5 to 8 are inherited from V. Therefore S is itself a vector space. □

Example 4.8 The following are examples of subspaces.

1. Let $\mathbf{x}$ be a nonzero vector in a vector space V. Then the set

$$S = \{\alpha\mathbf{x} : \alpha \in \mathbb{R}\}$$

  is a subspace of V. In fact, the linear span of any nonempty subset of V is a subspace of V.
2. The set $P(n)$ of all n-degree polynomials defined in Example 4.2 is a subspace of the vector space $F(S, \mathbb{R})$ defined in Example 4.5 with $S = \mathbb{R}$.
3. Let S be the vector space of all sequences in $\mathbb{R}$ defined in Example 4.3. Then the set of all convergent sequences in $\mathbb{R}$ is a subspace of S.

Let A and B be any nonempty subsets of a vector space V. Then the **Minkowski sum**, or simply the **sum** of sets A and B is defined as

$$A + B = \{\mathbf{a} + \mathbf{b} : \mathbf{a} \in A, \mathbf{b} \in B\}.$$

The definition can be extended to any number of subsets.

Example 4.9 In Example 2.9, Chap. 2 we define the production set of a firm. Suppose that an economy has n commodities and l firms. The firms' production sets are represented by $Y_j \subseteq \mathbb{R}^n$, $j = 1, \ldots, l$. In the analysis of the whole economy, we can add up the production of all firms into an aggregate firm, with the aggregate production set

$$Y = Y_1 + Y_2 + \cdots + Y_l = \sum_{j=1}^{l} Y_j.$$

A subset $A = \{x_1, x_2, \ldots, x_n\}$ of a vector space V is said to be **linearly dependent** if there exists a set of scalars $\{\alpha_1, \alpha_2, \ldots, \alpha_n\}$, not all zeros, such that

$$\alpha_1 x_1 + \alpha_2 x_2 + \cdots + \alpha_n x_n = \mathbf{0}. \tag{4.5}$$

That is, the zero vector is a linear combination of A with some nonzero scalars. Otherwise, the set A is called **linearly independent**.

Alternatively, we can define a set of vectors $A = \{x_1, x_2, \ldots, x_n\}$ to be linearly independent if the condition in Eq. (4.5) implies that $\alpha_i = 0$ for $i = 1, 2, \ldots, n$. You can verify the following consequences readily from the definitions:

1. If a set A contains the zero vector, then it is linearly dependent.
2. Any set containing a linearly dependent set is linearly dependent.
3. Any subset of a linearly independent set is linearly independent.

Let B be a linearly independent set of vectors of a vector space V. Then B is called a **basis** for V if span$(B) = V$. If B is finite, we call the cardinality of B the **dimension** of V. For example, if $|B| = n$, we write dim $V = n$.

If B is a basis for V, any vector $x \in V$ can be expressed as a linear combination of the vectors in B, that is,

$$x = \alpha_1 x_1 + \alpha_2 x_2 + \cdots + \alpha_n x_n,$$

for some scalars α_i, $i = 1, 2, \ldots, n$. We define the n-tuple $(\alpha_1, \alpha_2, \ldots, \alpha_n)$ formed by the scalars of the linear combination the **coordinates** of x relative to the basis B.

Example 4.10 Consider the vectors $(1, 2)$ and $(2, 1)$ in $\mathbb{R}^2$. If

$$\alpha(1, 2) + \beta(2, 1) = (0, 0),$$

then

$$\alpha + 2\beta = 0,$$
$$2\alpha + \beta = 0,$$

which gives $\alpha = \beta = 0$. Therefore the two vectors are linearly independent. For any vector $(x_1, x_2) \in \mathbb{R}^2$, we can write

$$(x_1, x_2) = \alpha(1, 2) + \beta(2, 1),$$

which implies that

$$\alpha + 2\beta = x_1,$$
$$2\alpha + \beta = x_2.$$

Solving the system of equation gives

$$\alpha = \frac{-x_1 + 2x_2}{3}, \quad \beta = \frac{2x_1 - x_2}{3}.$$

The two vectors span $\mathbb{R}^2$ and therefore form a basis for the vector space.

Example 4.11 Consider the following vectors in $\mathbb{R}^n$:

$$\mathbf{e}_1 = (1, 0, 0, \ldots, 0),$$

$$\mathbf{e}_2 = (0, 1, 0, \ldots, 0),$$

$$\vdots$$

$$\mathbf{e}_n = (0, 0, 0, \ldots, 1).$$

The set $B = \{\mathbf{e}_1, \mathbf{e}_2, \ldots, \mathbf{e}_n\}$ is a basis for $\mathbb{R}^n$. Let $\mathbf{x} = (x_1, x_2, \ldots, x_n)$ be any vector in $\mathbb{R}^n$. Then

$$\mathbf{x} = x_1\mathbf{e}_1 + x_2\mathbf{e}_2 + \cdots + x_n\mathbf{e}_n.$$

Therefore $(x_1, x_2, \ldots, x_n)$ is the coordinates of $\mathbf{x}$ relative to the basis B. In this case, the components of $\mathbf{x}$ coincide with the coordinates. For this reason we call $\{\mathbf{e}_1, \mathbf{e}_2, \ldots, \mathbf{e}_n\}$ the **standard basis** of $\mathbb{R}^n$.

It is clear from Examples 4.10 and 4.11 that the coordinates of a vector change with the choice of the basis for a vector space. We summarize some important facts on the basis and dimension of a vector space as follows.

1. If B is a finite set of linearly independent vectors that spans the vector space V, then any other basis for V must have the same cardinality as B. That is, dim V is a unique number.
2. If dim $V = n$, then any set containing more than n vectors in V is linearly dependent. Any set containing less than n vectors cannot span V.
3. If dim $V = n$ and W is a proper subspace of V, then dim $W < n$.
4. The set $\{0\}$ is called the zero subspace of V. But by definition the zero subspace is not linearly independent. Therefore we define $\dim\{0\} = 0$.

Example 4.12 You can verify the dimensions of the vector spaces in Examples 4.1 to 4.5 by identifying their standard basis:

- dim $\mathbb{R}^n = n$,
- dim $P(n) = n + 1$,
- S is infinite-dimensional,
- dim $\mathbb{M}^{m \times n} = m \times n$,
- $F(S, \mathbb{R})$ is infinite-dimensional.

In Example 4.7 we show that the set $B = \{f_0, f_1, \ldots, f_n\}$ where

$$f_i(x) = x^i, \quad i = 0, 1, \ldots, n,$$

is a basis for the vector space $P(n)$ of all n-degree polynomials. In this case the order of the set B is somewhat important because polynomials are expressed in increasing power of x. For this reason we define an **ordered basis** as an n-tuple of independent vectors in the desirable order, $B = (f_0, f_1, \ldots, f_n)$. Similarly, the standard ordered basis for $\mathbb{R}^n$ is $B = (e_1, e_2, \ldots, e_n)$.

Example 4.13 In matrix analysis, vectors in $\mathbb{R}^n$ are written as $\mathbb{M}^{n \times 1}$ column matrices to conform with matrix multiplication. Let A be an $n \times n$ invertible matrix. The column vectors of A, denoted by $a_1, a_2, \ldots, a_n \in \mathbb{R}^n$, form a basis for $\mathbb{R}^n$. This is because for any vector x in $\mathbb{R}^n$, we can define a vector $\alpha = (\alpha_1, \alpha_2, \ldots, \alpha_n)$ such that, written in matrix form, $x = A\alpha$, or

$$x = \alpha_1 a_1 + \alpha_2 a_2 + \cdots + \alpha_n a_n.$$

Therefore x can be expressed as a linear combination of $\{a_1, a_2, \ldots, a_n\}$. The coordinate α of x can be solved by the inverse of A, that is,

$$\alpha = A^{-1}x. \tag{4.6}$$

In Example 4.10,

$$A = \begin{pmatrix} 1 & 2 \\ 2 & 1 \end{pmatrix}.$$

The readers can verify the values of α and β using Eq. (4.6).

4.3   Linear Transformations

4.3.1   Introduction

A **linear transformation** is a function f which maps a vector space V into a vector space W such that for all $x, y \in V$ and scalar $\alpha \in \mathbb{R}$,

$$f(\alpha x + y) = \alpha f(x) + f(y).$$

That is, f is additive and linearly homogeneous. If f maps a vector space V into itself, it is called a **linear operator**.

The pre-image of the zero vector in W, $f^{-1}(0)$ is called the **kernel** or **null space** of f. It can be shown that the kernel of f is a subspace of V. The dimension of the kernel of f is called the **nullity** of f. Also, the range of f, $f(V)$ is a subspace of W. If V is finite-dimensional, the dimension of $f(V)$ is called the **rank** of f. The following theorem relates the rank and nullity of f.

Theorem 4.3 (Dimension Theorem) *Let $f : V \to W$ be a linear transformation. Suppose that V is finite-dimensional. Then*

$$rank\ f + nullity\ f = \dim V.$$

A proof of this important result can be found in Hoffman and Kunze (1971, 71–72). A transformation f is said to have **full rank** if

$$rank\ f = \min\{\dim V, \dim W\}.$$

The set of all linear transformations from V into W can be denoted by $L(V, W)$. For every $f, g \in L(V, W)$ and $\alpha \in \mathbb{R}$, define

1. $(f + g)(\mathbf{x}) = f(\mathbf{x}) + g(\mathbf{x})$,
2. $(\alpha f)(\mathbf{x}) = \alpha(f(\mathbf{x}))$.

It is straightforward to show that $f + g$ and αf are linear transformations. In fact $L(V, W)$ is a vector space, with $\dim L(V, W) = \dim V \times \dim W$.

4.3.2 Matrix Representations of Transformations

Let V be an n-dimensional vector space and W a m-dimensional vector space. Suppose that

$$\mathcal{B}_V = (\mathbf{v}_1, \ldots, \mathbf{v}_n)$$

is an ordered basis of V and

$$\mathcal{B}_W = (\mathbf{w}_1, \ldots, \mathbf{w}_m)$$

is an ordered basis of W. Let f be a linear transformation from V into W. Then f can be defined by its action on the vectors $\mathbf{v}_j$. In particular, $f(\mathbf{v}_j)$ can be expressed as a linear combination of the vectors in $\mathcal{B}_W$, that is,

$$f(\mathbf{v}_j) = a_{1j}\mathbf{w}_1 + a_{2j}\mathbf{w}_2 + \cdots + a_{mj}\mathbf{w}_m,$$

where $(a_{1j}, \ldots, a_{mj})$ are the coordinates of $f(\mathbf{v}_j)$ relative to the basis $\mathcal{B}_W$. For $j = 1, \ldots, n$, the $m \times n$ scalars a_{ij} can be arranged in the form of a matrix

$$A = \begin{pmatrix} a_{11} & a_{12} & \cdots & a_{1n} \\ a_{21} & a_{22} & \cdots & a_{2n} \\ \vdots & \vdots & & \vdots \\ a_{m1} & a_{m2} & \cdots & a_{mn} \end{pmatrix}.$$

That is, we write the coordinates of $f(\mathbf{v}_j)$ as the j-th column in A. A is called the **matrix representation** of f relative to the ordered bases $\mathcal{B}_V$ and $\mathcal{B}_W$. Since $\mathcal{B}_V$ is a basis of V, any vector $\mathbf{x} \in V$ can be expressed as a linear combination

$$\mathbf{x} = x_1 \mathbf{v}_1 + \cdots + x_n \mathbf{v}_n,$$

where $(x_1, \ldots, x_n)$ are the coordinates of $\mathbf{x}$ relative to $\mathcal{B}_V$. Then

$$\begin{aligned} f(\mathbf{x}) &= f(x_1 \mathbf{v}_1 + \cdots + x_n \mathbf{v}_n) \\ &= x_1 f(\mathbf{v}_1) + \cdots + x_n f(\mathbf{v}_n) \\ &= x_1 \sum_{i=1}^{m} a_{i1} \mathbf{w}_i + \cdots + x_n \sum_{i=1}^{m} a_{in} \mathbf{w}_i \\ &= \sum_{j=1}^{n} x_j \sum_{i=1}^{m} a_{ij} \mathbf{w}_i \\ &= \sum_{i=1}^{m} \sum_{j=1}^{n} a_{ij} x_j \mathbf{w}_i \\ &= \sum_{i=1}^{m} y_i \mathbf{w}_i \end{aligned}$$

where in the last line we define $y_i = \sum_{j=1}^{n} a_{ij} x_j$ for $i = 1, \ldots, m$. In matrix form we have

$$\mathbf{y} = A\mathbf{x}.$$

In fact $\mathbf{y} \in \mathbb{R}^m$ is the coordinates of $f(\mathbf{x})$ relative to the basis $\mathcal{B}_W$. In the above matrix form $\mathbf{x}$ and $\mathbf{y}$ are written in column matrix form with their coordinates as entries, that is,

$$\mathbf{x} = \begin{pmatrix} x_1 \\ x_2 \\ \vdots \\ x_n \end{pmatrix}, \qquad \mathbf{y} = \begin{pmatrix} y_1 \\ y_2 \\ \vdots \\ y_m \end{pmatrix}.$$

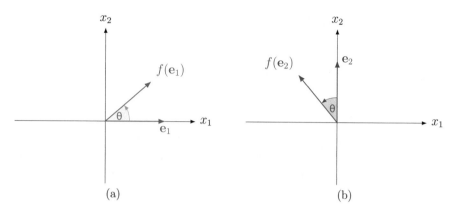

Fig. 4.2 The rotation transformation. (**a**) Action on e_1. (**b**) Action on e_2

You should verify that the dimensions of the matrices are compatible in the multiplication.

It is not difficult to see that if A and B are the matrix representations of f and g with respect to the same bases, then $A + B$ and αA are the respective matrix representations of $f + g$ and αf. For the composite transformation $g \circ f$ in Exercise 18, the matrix representation is BA.

Example 4.14 Suppose that $f : \mathbb{R}^2 \to \mathbb{R}^2$ is a linear transformation that rotates a vector by an angle θ in the counter-clockwise direction. Let $\mathcal{B} = (e_1, e_2)$ be the standard basis in $\mathbb{R}^2$. Figure 4.2 depicts the actions of f on e_1 and e_2. The coordinates of their images are $(\cos\theta, \sin\theta)$ and $(-\sin\theta, \cos\theta)$ respectively. Therefore the matrix representation of f relative to $\mathcal{B}$ is

$$A = \begin{pmatrix} \cos\theta & -\sin\theta \\ \sin\theta & \cos\theta \end{pmatrix}.$$

4.3.3   Linear Functionals

When the co-domain of f is $\mathbb{R}$, the transformation becomes a linear functional. Linear functionals appear frequently in economic models. For example, in production theory, the profit of a firm with net output vector $\mathbf{y} = (y_1, \ldots, y_n) \in \mathbb{R}^n$ is a linear functional

$$f(\mathbf{y}) = p_1 y_1 + \cdots + p_n y_n,$$

where p_i is the market price of good i, $i = 1, \ldots, n$. The following are some other important examples.

Example 4.15 Let A be an $n \times n$ square matrix. The **trace** of A is defined as

$$\text{tr}(A) = a_{11} + a_{22} + \cdots + a_{nn}.$$

Example 4.16 The mathematical expectation or expected value of a random variable X with a probability density function $f(x)$ is defined as

$$E[X] = \int_{-\infty}^{\infty} x f(x) \, dx.$$

Example 4.17 Let $[a, b]$ be an interval in $\mathbb{R}$ and let $\mathscr{C}([a, b])$ be the set of all continuous functions on $[a, b]$. Then for every $g \in \mathscr{C}([a, b])$, the integral of g,

$$L(g) = \int_{a}^{b} g(t) \, dt$$

defines a linear functional L on $\mathscr{C}([a, b])$.

If V has a finite dimension n, the matrix representation of a linear functional f relative to an ordered basis $\mathcal{B}_V$ in V is

$$A = [a_1 \ a_2 \ \cdots \ a_n] \tag{4.7}$$

and

$$f(\mathbf{x}) = a_1 x_1 + a_2 x_2 + \cdots + a_n x_n,$$

where $(x_1, \ldots, x_n)$ are the coordinates of $\mathbf{x}$ relative to $\mathcal{B}_V$. The collection of all linear functionals on V, $L(V, \mathbb{R})$ is a vector space. This vector space V^* is often called the **dual space** of V. It is clear from (4.7) that $\dim V^* = \dim V = n$.

By definition the rank of a linear functional is 1. Therefore by the Dimension Theorem

$$\text{nullity } f = \dim V - \text{rank } f = n - 1.$$

The kernel of f is called a **hyperspace** of V. Similarly, a **hyperplane** of f at the level c is defined as[1]

$$H_f(c) = \{\mathbf{x} \in V : f(\mathbf{x}) = c\}. \tag{4.8}$$

For example,

---

[1] Recall that in functional analysis a hyperplane is the contour of f at c.

$$H_f(\pi) = \{\mathbf{y} \in \mathbb{R}^n : p_1 y_1 + \cdots + p_n y_n = \pi\} \tag{4.9}$$

is an isoprofit hyperplane in production analysis. The set of $\mathbf{y}$ in $H_f(\pi)$ gives the same profit level π. Similarly,

$$H_f(M) = \{\mathbf{x} \in \mathbb{R}^n : p_1 x_1 + \cdots + p_n x_n = M\}$$

defines the budget constraint in consumer analysis with prices $\mathbf{p} = (p_1, \ldots, p_n)$ and total expenditure M. Notice that $H_f(c)$ is not a subspace of V unless $c = 0$.

4.3.4 Multilinear Transformations

Let V be a vector space. Then a function f which maps the product space $V^r = V \times V \times \cdots \times V$ to a vector space W is **multilinear** if $f(\mathbf{x}_1, \mathbf{x}_2, \ldots, \mathbf{x}_r)$ is linear in each of the $\mathbf{x}_i$ when the other $\mathbf{x}_j$'s are held fixed. That is, for all $\mathbf{x}_i, \mathbf{y}_i \in V, \alpha \in \mathbb{R}$ and $i = 1, \ldots, r$,

$$f(\mathbf{x}_1, \ldots, \alpha \mathbf{x}_i + \mathbf{y}_i, \ldots, \mathbf{x}_r) = \alpha f(\mathbf{x}_1, \ldots, \mathbf{x}_i, \ldots, \mathbf{x}_r) + f(\mathbf{x}_1, \ldots, \mathbf{y}_i, \ldots, \mathbf{x}_r).$$

The function f above is often called r-linear. When $r = 2$, f is called **bilinear**. For example, if $\dim V = n$, the matrix representation of a bilinear functional can be expressed as

$$f(\mathbf{x}, \mathbf{y}) = \mathbf{x}^T A \mathbf{y},$$

where A is an $n \times n$ square matrix.

4.4 Inner Product Spaces

Let V be a vector space. An **inner product** on V is a function which assigns an ordered pair of vectors $\mathbf{x}, \mathbf{y} \in V$ a real number, denoted by $\mathbf{x}^T \mathbf{y}$, which satisfies the following axioms: For all $\mathbf{x}, \mathbf{y}, \mathbf{z} \in V$ and $\alpha \in \mathbb{R}$,

1. $(\mathbf{x} + \mathbf{y})^T \mathbf{z} = \mathbf{x}^T \mathbf{z} + \mathbf{y}^T \mathbf{z}$,
2. $(\alpha \mathbf{x})^T \mathbf{y} = \alpha(\mathbf{x}^T \mathbf{y})$,
3. $\mathbf{x}^T \mathbf{y} = \mathbf{y}^T \mathbf{x}$,
4. $\mathbf{x}^T \mathbf{x} \geq 0$, $\mathbf{x}^T \mathbf{x} = 0$ implies that $\mathbf{x} = \mathbf{0}$.

Inner products are often denoted by $\mathbf{x} \cdot \mathbf{y}$, $\langle \mathbf{x}, \mathbf{y} \rangle$ or $(\mathbf{x}|\mathbf{y})$.

Example 4.18 When $V = \mathbb{R}^n$, the Euclidean space of dimension n, the dot product of two vector is defined as

$$\mathbf{x}^{\mathrm{T}}\mathbf{y} = x_1 y_1 + \cdots + x_n y_n.$$

Example 4.19 In $\mathbb{R}^2$, define $\mathbf{x}^{\mathrm{T}}\mathbf{y}$ as

$$\mathbf{x}^{\mathrm{T}}\mathbf{y} = x_1 y_1 - x_2 y_1 - x_1 y_2 + 4x_2 y_2.$$

It is straightforward to show that the above is an inner product.

Example 4.20 Let $\mathscr{C}([0, 1])$ be the set of all continuous functions on $[0, 1]$. For any $f, g \in \mathscr{C}([0, 1])$, an inner product can be defined as

$$f^{\mathrm{T}}g = \int_0^1 f(t)g(t)dt.$$

The study of inner products is concerned with concepts of the "length" of a vector and the "angle" between two vectors. The first idea involves the definition of the **norm** of a vector as follows: For all $\mathbf{x} \in V$,

$$\|\mathbf{x}\| = (\mathbf{x}^{\mathrm{T}}\mathbf{x})^{1/2}.$$

The above definition of course satisfies the three axioms of a **normed vector space**, that is, for all $\mathbf{x}, \mathbf{y} \in V$ and $\alpha \in \mathbb{R}$,

$$\|\mathbf{x}\| \geq 0, \ \|\mathbf{0}\| = 0, \tag{4.10}$$

$$\|\alpha\mathbf{x}\| = |\alpha| \, \|\mathbf{x}\|, \tag{4.11}$$

$$\|\mathbf{x} + \mathbf{y}\| \leq \|\mathbf{x}\| + \|\mathbf{y}\|. \tag{4.12}$$

Axioms (4.10) and (4.11) are straightforward to verify. Axiom (4.12) requires the application of the Cauchy-Schwarz Inequality and will be presented in Sect. 4.5 below.

The following definitions are concerned with vectors being "perpendicular" to each other. Let $\mathbf{x}, \mathbf{y} \in V$. The **angle** θ between $\mathbf{x}$ and $\mathbf{y}$ is implicitly defined as

$$\cos \theta = \frac{\mathbf{x}^{\mathrm{T}}\mathbf{y}}{\|\mathbf{x}\|\|\mathbf{y}\|}.$$

It follows that θ is $\pi/2$ or $90°$ when the inner product of $\mathbf{x}$ and $\mathbf{y}$ is zero. Formally, $\mathbf{x}$ and $\mathbf{y}$ are **orthogonal** to each other, sometimes written as $\mathbf{x} \perp \mathbf{y}$, if $\mathbf{x}^{\mathrm{T}}\mathbf{y} = 0$. A set of vectors $S \subseteq V$ is called an **orthogonal set** if all pairs of distinct vectors in S are orthogonal. Moreover, if every vector in S has norm equals to 1, S is called an **orthonormal set**. The **orthogonal complement** of any set $S \subseteq V$ is defined as all the vectors in V which are orthogonal to every vector in S, that is,

Fig. 4.3 An orthogonal projection from $\mathbb{R}^3$ onto $\mathbb{R}^2$

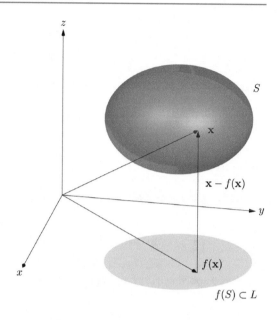

$$S^{\perp} = \{\mathbf{x} \in V : \mathbf{x}^{\mathrm{T}}\mathbf{y} = 0, \mathbf{y} \in S\}.$$

An $n \times n$ square matrix A is called orthogonal if the rows of A form a orthonormal set in $\mathbb{R}^n$. Orthogonal matrices appear frequently in econometric analysis.

The **orthogonal projection** of a set $S \subseteq V$ onto a subspace L is a linear transformation $f : S \to L$ such that for every $\mathbf{x} \in S$, $(\mathbf{x} - f(\mathbf{x})) \in L^{\perp}$. Figure 4.3 illustrates the orthogonal projection of a set S in $\mathbb{R}^3$ onto the subspace of the x-y plane. The range of the projection, $f(S)$, is the "shadow" of S on the x-y plane. For any $\mathbf{x} \in S$, the vector $\mathbf{x} - f(\mathbf{x})$ is parallel to the z axis and therefore orthogonal to the x-y plane.

4.5 Inequalities

4.5.1 Cauchy-Schwarz Inequality

Let V be an inner product space. For all $\mathbf{x} \in V$, define the norm as $\|\mathbf{x}\| = (\mathbf{x}^{\mathrm{T}}\mathbf{x})^{1/2}$. Then for all $\mathbf{x}, \mathbf{y} \in V$, $|\mathbf{x}^{\mathrm{T}}\mathbf{y}| \leq \|\mathbf{x}\| \, \|\mathbf{y}\|$.

Proof The result is trivial if $\mathbf{y} = \mathbf{0}$. So assume that $\mathbf{y} \neq \mathbf{0}$. Then for any $\alpha \in \mathbb{R}$, we have

$$0 \leq \|\mathbf{x} - \alpha\mathbf{y}\|^2$$
$$= (\mathbf{x} - \alpha\mathbf{y})^{\mathrm{T}}(\mathbf{x} - \alpha\mathbf{y})$$
$$= \mathbf{x}^{\mathrm{T}}(\mathbf{x} - \alpha\mathbf{y}) - \alpha\mathbf{y}^{\mathrm{T}}(\mathbf{x} - \alpha\mathbf{y})$$

$$= \mathbf{x}^T\mathbf{x} - \alpha\mathbf{x}^T\mathbf{y} - \alpha\mathbf{y}^T\mathbf{x} + \alpha^2\mathbf{y}^T\mathbf{y}$$
$$= \mathbf{x}^T\mathbf{x} - 2\alpha\mathbf{x}^T\mathbf{y} + \alpha^2\mathbf{y}^T\mathbf{y}.$$

Setting $\alpha = \mathbf{x}^T\mathbf{y}/\mathbf{y}^T\mathbf{y}$, the inequality becomes

$$0 \le \mathbf{x}^T\mathbf{x} - \frac{(\mathbf{x}^T\mathbf{y})^2}{\mathbf{y}^T\mathbf{y}} = \|\mathbf{x}\|^2 - \frac{(\mathbf{x}^T\mathbf{y})^2}{\|\mathbf{y}\|^2},$$

and the result follows.   □

Applying the Cauchy-Schwarz inequality to Example 4.20 above, we obtain

$$\left| \int_0^1 f(t)g(t)dt \right| \le \left(\int_0^1 |f(t)|^2 dt \right)^{1/2} \left(\int_0^1 |g(t)|^2 dt \right)^{1/2},$$

which is a special case of an important result in functional analysis called Hölder's Inequality.

4.5.2   Triangular Inequality

For all $\mathbf{x}, \mathbf{y}$ in a normed vector space V,

$$\|\mathbf{x} + \mathbf{y}\| \le \|\mathbf{x}\| + \|\mathbf{y}\|.$$

Proof For $\mathbf{x}, \mathbf{y} \in V$, we have

$$\begin{aligned}
\|\mathbf{x} + \mathbf{y}\|^2 &= (\mathbf{x} + \mathbf{y})^T(\mathbf{x} + \mathbf{y}) \\
&= \mathbf{x}^T\mathbf{x} + \mathbf{x}^T\mathbf{y} + \mathbf{y}^T\mathbf{x} + \mathbf{y}^T\mathbf{y} \\
&= \|\mathbf{x}\|^2 + 2\mathbf{x}^T\mathbf{y} + \|\mathbf{y}\|^2 \\
&\le \|\mathbf{x}\|^2 + 2|\mathbf{x}^T\mathbf{y}| + \|\mathbf{y}\|^2 \\
&\le \|\mathbf{x}\|^2 + 2\|\mathbf{x}\|\,\|\mathbf{y}\| + \|\mathbf{y}\|^2 \\
&= (\|\mathbf{x}\| + \|\mathbf{y}\|)^2
\end{aligned}$$

and the result follows. Note that in the second inequality above we have applied the Cauchy-Schwarz Inequality.   □

4.6   Determinants

Let $L(V)$ be the collection of all linear operators on a finite-dimensional vector space V. Our goal is to find a necessary and sufficient condition for any f in $L(V)$ to have an inverse operator. That is, we want to know under what condition f is a bijection. It turns out that a functional on $L(V)$ can do the trick. The functional, however, is more conveniently defined using the matrix representations of $L(V)$ relative to a basis $\mathcal{B}_V$. Therefore we define a functional D which assigns each $n \times n$ matrix A a real number and satisfies the following properties:

1. D is multilinear (recall Sect. 4.3.4) in the row vectors of A. That is, if we let $\mathbf{x}_i$ be the i-th row of A, $i = 1, \ldots, n$ so that we can write

$$D(A) = D(\mathbf{x}_1, \ldots, \mathbf{x}_n),$$

then

$$D(\mathbf{x}_1, \ldots, \alpha\mathbf{x}_i + \mathbf{y}_i, \ldots, \mathbf{x}_n) = \alpha D(\mathbf{x}_1, \ldots, \mathbf{x}_i, \ldots, \mathbf{x}_n) + D(\mathbf{x}_1, \ldots, \mathbf{y}_i, \ldots, \mathbf{x}_n).$$

2. If two rows of A are identical, then $D(A) = 0$.
3. $D(I) = 1$. That is, the value of the identity matrix is 1.

Any functional on $L(V)$ which satisfies the above three conditions is called a **determinant function**. It turns out that *the* determinant function is unique.[2] The determinant of A is often written as det A or $|A|$.

The formula for the determinant of A can be expressed inductively as follows. The determinant of a 1×1 matrix $A = (a)$ is simply a. For $n > 1$, define the **minor**, A_{ij}, of the element a_{ij} of A as the determinant of the $(n-1) \times (n-1)$ matrix by deleting the i-th row and j-th column of A. The **determinant** of A is given by

$$|A| = \sum_{j=1}^{n} (-1)^{i+j} a_{ij} A_{ij}. \tag{4.13}$$

Notice that in (4.13) the row i is not specified. That means we can find the determinant by choosing any row. In fact we can interchange the indices i and j and evaluate $|A|$ along any column instead.

Some important properties of the determinant function are given below:

1. $|AB| = |A||B|$.
2. $|A^{\mathrm{T}}| = |A|$.
3. If A has a row (column) of zeros, then $|A| = 0$.
4. If B is obtained from A by interchanging two rows (columns) of A, then $|B| = -|A|$.
5. If B is obtained from A by adding the multiple of one row (column) of A to another row (column), then $|B| = |A|$.

---

[2] See Hoffman and Kunze (1971, Chapter 5).

6. If B is obtained from A by multiplying a row (column) of A by a scalar α, then
   $|B| = \alpha |A|$.
7. Suppose

$$A = \begin{pmatrix} A_{11} & A_{12} \\ \mathbf{0} & A_{22} \end{pmatrix}$$

where A_{11} and A_{22} are square matrices and $\mathbf{0}$ represents a zero matrix with conforming dimension. Then

$$|A| = |A_{11}| \, |A_{22}|.$$

4.7   Inverse Transformations

4.7.1   Systems of Linear Equations

Let f be a linear operator on an n-dimensional vector space V. For a given vector $\mathbf{b} \in V$, we are interested in the pre-image $f^{-1}(\mathbf{b})$. In particular, we want to know if we can find a unique $\mathbf{x} \in V$ such that $f(\mathbf{x}) = \mathbf{b}$. In matrix form we have

$$A\mathbf{x} = \mathbf{b} \tag{4.14}$$

where

$$A = \begin{pmatrix} a_{11} & a_{12} & \cdots & a_{1n} \\ a_{21} & a_{22} & \cdots & a_{2n} \\ \vdots & \vdots & & \vdots \\ a_{n1} & a_{n2} & \cdots & a_{nn} \end{pmatrix}$$

is the matrix representation of f relative to the standard basis of V, and the coordinates of $\mathbf{x}$ and $\mathbf{b}$ are written as column vectors:

$$\mathbf{x} = \begin{pmatrix} x_1 \\ x_2 \\ \vdots \\ x_n \end{pmatrix}, \quad \mathbf{b} = \begin{pmatrix} b_1 \\ b_2 \\ \vdots \\ b_n \end{pmatrix}.$$

Of course if f is a bijection, f^{-1} is the inverse linear operator such that $f^{-1} \circ f$ becomes the identity operator. In matrix form the **inverse** of A is denoted by A^{-1} and

$$A^{-1}A = AA^{-1} = I \tag{4.15}$$

where I is the n-dimensional identity matrix.[3] Pre-multiply both sides of (4.14) by A^{-1} gives

$$\mathbf{x} = A^{-1}\mathbf{b}.$$

4.7.2 Inverse of a Matrix

Recall in Eq. (4.13) that the minor, A_{ij}, of the element a_{ij} of A is defined as the determinant of the $(n-1) \times (n-1)$ matrix by deleting the i-th row and j-th column of A. The **cofactor** of a_{ij}, C_{ij}, is defined as $(-1)^{i+j} A_{ij}$. Using the cofactors, $|A|$ in (4.13) can be written as

$$|A| = \sum_{j=1}^{n} a_{ij} C_{ij}. \tag{4.16}$$

Now we show that if we replace the row $a_{i1}, a_{i2}, \ldots, a_{in}$ in (4.16) with another row k in A, then

$$\sum_{j=1}^{n} a_{kj} C_{ij} = 0. \tag{4.17}$$

To see this, let B be the matrix obtained from A by replacing the i-row with the k-row. Then B has two identical rows and therefore $|B| = 0$. The minors of the i-th row of B, however, are still the same as those of A. That is, $B_{ij} = A_{ij}$ for $j = 1, \ldots, n$. Now we have

$$\sum_{j=1}^{n} a_{kj} C_{ij} = \sum_{j=1}^{n} (-1)^{i+j} a_{kj} A_{ij}$$

$$= \sum_{j=1}^{n} (-1)^{i+j} b_{ij} B_{ij}$$

$$= |B| = 0.$$

Equations (4.16) and (4.17) can be summarized as

$$\sum_{j=1}^{n} a_{kj} C_{ij} = \begin{cases} |A| & \text{if } i = k, \\ 0 & \text{otherwise.} \end{cases} \tag{4.18}$$

---

[3] See Exercise 34 at the end of the chapter.

The transpose of the $n \times n$ matrix of cofactors of A is called the **adjoint** of A, or adj A. Then (4.18) can be expressed as

$$A(\text{adj } A) = |A|I. \tag{4.19}$$

It is clear from (4.19) that if $|A| \neq 0$, then the inverse of A defined in (4.15) is

$$A^{-1} = \frac{1}{|A|}(\text{adj } A) = \frac{1}{|A|}C^{\text{T}},$$

where C is the n-dimensional square matrix with elements C_{ij}, the cofactor of a_{ij}. If $|A| = 0$, then the linear operator f is not a bijection and the inverse operator does not exist. In this case f and A are called **singular**.

In the simplest case of $n = 2$, let

$$A = \begin{pmatrix} a & b \\ c & d \end{pmatrix}.$$

In order for A to have an inverse, we need $|A| = ad - bc \neq 0$. The minor of a is d, the minor of b is c, etc. Then $C_{11} = (-1)^{1+1}d = d$ and $C_{12} = (-1)^{1+2}c = -c$, etc. Putting everything together, we have

$$A^{-1} = \frac{1}{ad - bc} \begin{pmatrix} d & -b \\ -c & a \end{pmatrix}.$$

4.7.3   Cramer's Rule

Instead of solving for the solution of the whole system of linear equations in (4.14), sometimes we need the solution for a single component, x_i. In this case we do not have to find A^{-1}, which is a very time consuming procedure if done manually. The Cramer's rule provides a shortcut to do just that. Suppose we want the solution of x_i in (4.14). We proceed as follows.

1. Find the determinant of A. Ensure that $|A| \neq 0$.
2. Replace the i-th column of A by the column vector $\mathbf{b}$. Call this modified matrix A_i. In other words,

$$A_i = \begin{pmatrix} a_{11} & \cdots & b_1 & \cdots & a_{1n} \\ a_{21} & \cdots & b_2 & \cdots & a_{2n} \\ \vdots & & \vdots & & \vdots \\ a_{n1} & \cdots & b_n & \cdots & a_{nn} \end{pmatrix}$$

3. Find $|A_i|$.

4. The solution is

$$x_i = \frac{|A_i|}{|A|}.$$

4.7.4 Properties of Inverses

Some useful properties of the inverses:

1. If A and B are invertible, then $(AB)^{-1} = B^{-1}A^{-1}$.
2. $(A^{-1})^{-1} = A$.
3. $(A^{\mathrm{T}})^{-1} = (A^{-1})^{\mathrm{T}}$.
4. $|A^{-1}| = 1/|A|$.
5. The inverse of an upper (lower) triangular matrix is upper (lower) triangular.
6. Suppose

$$A = \begin{pmatrix} A_{11} & \mathbf{0} \\ \mathbf{0} & A_{22} \end{pmatrix}$$

where A_{11} and A_{22} are square matrices and $\mathbf{0}$ represents a zero matrix with conforming dimension. Then

$$A^{-1} = \begin{pmatrix} A_{11}^{-1} & \mathbf{0} \\ \mathbf{0} & A_{22}^{-1} \end{pmatrix}$$

4.8 Eigenvectors and Eigenvalues of Linear Operators

Let f be a linear operator on an n-dimensional vector space V. The matrix representation of f relative to any basis of V is an $n \times n$ square matrix. In particular, we ask whether there exists an ordered basis $\mathcal{B} = (\mathbf{x}_1, \ldots, \mathbf{x}_n)$ such that the matrix representation of f, Λ, is diagonal. If such a basis exists, a lot of important properties of f such as the determinant, rank, the kernel and the definiteness of its quadratic form (see Sect. 4.9) can be read quickly from Λ. This happens if and only if $f(\mathbf{x}_i) = \lambda_i \mathbf{x}_i$ for $i = 1, \ldots, n$. This implies that the range of f is spanned by those $\mathbf{x}_i$'s for which $\lambda_i \neq 0$, and the kernel of f is spanned by the remaining $\mathbf{x}_i$'s. In this section we investigate under what conditions $\mathcal{B}$ exists and how it can be found.

4.8.1 Definitions

Let f be a linear operator on an n-dimensional vector space V. Then a nonzero vector $\mathbf{x} \in V$ is an **eigenvector** of f if

$$f(\mathbf{x}) = \lambda\mathbf{x}. \tag{4.20}$$

The scalar λ is called an **eigenvalue** or **characteristic value** of f. Note that if $\lambda = 1$, then $\mathbf{x}$ is a fixed point of f. The set of all the eigenvalues of f is called the **spectrum** of f. The largest of the absolute values of the eigenvalues is called the **spectral radius** of f. Multiplying both sides of (4.20) by $\alpha \neq 0$, we have

$$\alpha f(\mathbf{x}) = \alpha(\lambda\mathbf{x})$$

or

$$f(\alpha\mathbf{x}) = \lambda(\alpha\mathbf{x})$$

so that eigenvectors are unique up to scalar multiplication, that is, if $\mathbf{x}$ is an eigenvector of f, $\alpha\mathbf{x}$ is also an eigenvector for any $\alpha \neq 0$. Conventionally we express eigenvectors in their normalized form, that is, we choose $\alpha = 1/\|\mathbf{x}\|$ so that the resulting eigenvector has norm equal to one.

Let A be the matrix representation of f relative to the standard basis of V. Then definition (4.20) becomes

$$A\mathbf{x} = \lambda\mathbf{x}$$

so that

$$(A - \lambda I)\mathbf{x} = \mathbf{0}. \tag{4.21}$$

Since $\mathbf{x} \neq \mathbf{0}$, (4.21) implies that the matrix $A - \lambda I$ is singular so that $|A - \lambda I| = 0$. By expanding the determinant we get the **characteristic equation** for λ. This equation is a polynomial of degree n in λ. By the Fundamental Theorem in Algebra we know that it has n roots, which can be complex numbers. Therefore a linear operator on a n-dimensional vector space has n eigenvalues, not necessary distinct.

Example 4.21 Suppose that a linear operator f on $\mathbb{R}^2$ is represented by the matrix

$$A = \begin{pmatrix} 4 & 2 \\ 2 & 1 \end{pmatrix}$$

relative to the standard basis. Then

$$(A - \lambda I) = \begin{pmatrix} 4 - \lambda & 2 \\ 2 & 1 - \lambda \end{pmatrix}.$$

The characteristic equation is

$$\lambda^2 - 5\lambda = 0.$$

The roots of this quadratic equation are $\lambda_1 = 5$ and $\lambda_2 = 0$. For $\lambda_1 = 5$, (4.21) implies that

$$\begin{pmatrix} -1 & 2 \\ 2 & -4 \end{pmatrix} \begin{pmatrix} x_1 \\ x_2 \end{pmatrix} = \mathbf{0},$$

so that $x_1 = 2x_2$, or

$$\mathbf{x}_1 = \begin{pmatrix} x_1 \\ x_2 \end{pmatrix} = \alpha \begin{pmatrix} 2 \\ 1 \end{pmatrix}.$$

To normalize $\mathbf{x}_1$ we let $\alpha = 1/\sqrt{2^2 + 1^2} = 1/\sqrt{5}$ so that the normalized eigenvector is

$$\mathbf{x}_1 = \begin{pmatrix} 2/\sqrt{5} \\ 1/\sqrt{5} \end{pmatrix}.$$

Similarly, for $\lambda_2 = 0$, the normalized eigenvector is

$$\mathbf{x}_2 = \begin{pmatrix} 1/\sqrt{5} \\ -2/\sqrt{5} \end{pmatrix}.$$

We can observe some properties of f from the results. First, the determinant of A is zero since it has a zero eigenvalue. Second, the rank of f is 1. Third, $\mathcal{B} = \{\mathbf{x}_1, \mathbf{x}_2\}$ is an orthonormal basis for $\mathbb{R}^2$ and

$$P = (\mathbf{x}_1 \; \mathbf{x}_2) = \frac{1}{\sqrt{5}} \begin{pmatrix} 2 & 1 \\ 1 & -2 \end{pmatrix}$$

is an orthogonal matrix. Finally, if we form a diagonal matrix Λ with the eigenvalues, that is,

$$\Lambda = \begin{pmatrix} 5 & 0 \\ 0 & 0 \end{pmatrix},$$

then it is straightforward to show that $A = P\Lambda P^{\mathrm{T}}$.

4.8.2 Symmetric Operators

A linear operator f is **symmetric** if for all $\mathbf{x}, \mathbf{y} \in V$,

$$f(\mathbf{x})^{\mathrm{T}}\mathbf{y} = \mathbf{x}^{\mathrm{T}} f(\mathbf{y}).$$

In the matrix representation, this means that $\mathbf{x}^T A \mathbf{y} = \mathbf{y}^T A \mathbf{x}$. Since $\mathbf{x}^T A \mathbf{y} = \mathbf{y}^T A^T \mathbf{x}$ by transposition, symmetry requires that $A = A^T$. The eigenvalues and eigenvectors of a symmetric linear operator have many properties that are useful in econometrics and economics:

1. All the eigenvalues are real numbers.
2. If an eigenvalue has k multiple roots, it also has k orthogonal eigenvectors.
3. The n normalized eigenvectors form an orthogonal matrix. For example, if $\mathbf{x}_1, \mathbf{x}_2, \ldots, \mathbf{x}_n$ are the n normalized eigenvectors of A, then the square matrix $P = (\mathbf{x}_1\ \mathbf{x}_2\ \cdots\ \mathbf{x}_n)$ is orthogonal. That is, we have $P^T P = I$, or equivalently, $P^T = P^{-1}$.
4. Diagonal decomposition (Spectral Theorem):

$$A = P \Lambda P^T$$

   where $\Lambda = \mathrm{diag}(\lambda_1, \lambda_2, \ldots, \lambda_n)$, that is, Λ is an $n \times n$ diagonal matrix with the λ_i at the principal diagonal and the off-diagonal elements are all zero.
5. The sum of the eigenvalues of A (including any multiplicity) is equal to the sum of the diagonal elements (trace) of A, that is,

$$\sum_{i=1}^{n} \lambda_i = \sum_{i=1}^{n} a_{ii}.$$

6. The product of the eigenvalues of A is equal to its determinant, that is,

$$\prod_{i=1}^{n} \lambda_i = |A|.$$

7. The rank of A (rank f) is equal to the number of nonzero eigenvalues.
8. The eigenvalues of A^2 ($f \circ f$) are $\lambda_1^2, \lambda_2^2, \ldots, \lambda_n^2$, but the eigenvectors are the same as A.
9. If A is nonsingular, the eigenvalues of A^{-1} are $\lambda_1^{-1}, \lambda_2^{-1}, \ldots, \lambda_n^{-1}$, but the eigenvectors are the same as A.
10. A matrix A is **idempotent** if $A^2 = A$. The eigenvalues of an idempotent matrix are either 0 or 1.

4.9   Quadratic Forms

Suppose that f is a symmetric linear operator on a n-dimensional vector space V. A **quadratic form** is a functional $Q : V \to \mathbb{R}$ such that

$$Q(\mathbf{x}) = \mathbf{x}^T f(\mathbf{x}) = \mathbf{x}^T A \mathbf{x} = \sum_{i=1}^{n} \sum_{j=1}^{n} a_{ij} x_i x_j,$$

where A is the symmetric matrix representation of f.[4] The quadratic form is said to process some definite properties if it satisfies one of the following definitions: For all $\mathbf{x} \neq \mathbf{0}$,

1. A is **positive definite** if $\mathbf{x}^T A \mathbf{x} > 0$.
2. A is **positive semi-definite** if $\mathbf{x}^T A \mathbf{x} \geq 0$, with equality for some $\mathbf{x}$.
3. A is **negative definite** if $\mathbf{x}^T A \mathbf{x} < 0$.
4. A is **negative semi-definite** if $\mathbf{x}^T A \mathbf{x} \leq 0$, with equality for some $\mathbf{x}$.

The quadratic form is called **indefinite** if it does not satisfy any of the above properties.

4.9.1 Positive Definite Matrices

The following result shows two characterizations of positive definiteness. But first we define the principal matrix A_k of A, for $k = 1, 2, \ldots, n$ as the $k \times k$ submatrix formed by the first k rows and k columns of A. The determinant of A_k, $|A_k|$, is called the k-th principal minor of A.

Theorem 4.4 *The following three statements are equivalent:*

1. A is positive definite.
2. All eigenvalues of A are positive.
3. All principal minors of A are positive, that is, $|A_k| > 0$ for $k = 1, 2, \ldots, n$.

Statements 2 and 3 of Theorem 4.4 imply that a positive definite matrix is nonsingular. Testing for positive semi-definiteness is more involving. First let $K = \{1, 2, \ldots, k\}$ and Π be the set of all permutations on K. For example, for $k = 3$, Π consists $3! = 3 \times 2 \times 1 = 6$ permutations: $\pi_1 = \{1, 2, 3\}, \pi_2 = \{1, 3, 2\}, \pi_3 = \{2, 1, 3\}, \pi_4 = \{2, 3, 1\}, \pi_5 = \{3, 1, 2\}$ and $\pi_6 = \{3, 2, 1\}$. Now denote A_k^π as the $k \times k$ principal matrix of A with the rows and columns positions rearranged according to the permutation $\pi \in \Pi$. For example, the first two permutations of A_3 are:

$$A_3^{\pi_1} = \begin{pmatrix} a_{11} & a_{12} & a_{13} \\ a_{21} & a_{22} & a_{23} \\ a_{31} & a_{32} & a_{33} \end{pmatrix}$$

and

---

[4] Any asymmetric metric B can be converted into a symmetric matrix by taking $A = \frac{1}{2}(B + B^T)$. The values of the quadratic form will be the same.

$$A_3^{\pi_2} = \begin{pmatrix} a_{11} & a_{13} & a_{12} \\ a_{31} & a_{33} & a_{32} \\ a_{21} & a_{23} & a_{22} \end{pmatrix}.$$

Theorem 4.5 *The following three statements are equivalent:*

1. *A is positive semi-definite.*
2. *All eigenvalues of A are nonnegative, with at least one equal to zero.*
3. $|A_k^{\pi}| \geq 0$ *for* $k = 1, 2, \ldots, n$ *and for all* $\pi \in \Pi$.

4.9.2   Negative Definite Matrices

Testing for negative and negative semi-definite matrices is similar with a slight difference from the above.

Theorem 4.6 *The following three statements are equivalent:*

1. *A is negative definite.*
2. *All eigenvalues of A are negative.*
3. $(-1)^k |A_k| > 0$ *for* $k = 1, 2, \ldots, n$.

Theorem 4.7 *The following three statements are equivalent:*

1. *A is negative semi-definite.*
2. *All eigenvalues of A are nonpositive, with at least one equal to zero.*
3. $(-1)^k |A_k^{\pi}| \geq 0$ *for* $k = 1, 2, \ldots, n$ *and for all* $\pi \in \Pi$.

Notice, however, that the quadratic form $Q(\mathbf{x})$ is negative (semi-)definite if and only if $-Q(\mathbf{x})$ is positive (semi-)definite. Therefore if A is negative (semi-)definite, we can apply Theorems 4.4 and 4.5 to $-A$.

The following example illustrates why we need to examine the permutation matrices when checking for semi-definiteness. Suppose that

$$A = \begin{pmatrix} 0 & 0 & 0 \\ 0 & 1 & 0 \\ 0 & 0 & -1 \end{pmatrix}.$$

In this case $|A_1| = |A_2| = |A| = 0$. Therefore A does not satisfy the requirement of a positive or negative definite matrix. It does, however, satisfy the weak inequality of condition 3 in Theorem 4.5 and condition 3 in Theorem 4.7 without any permutation. We might want to conclude that A is both positive semi-definite and negative semi-definite. This turns out not to be the case because $\mathbf{e}_2^T A \mathbf{e}_2 = 1 > 0$

and $e_3^T A e_3 = -1 < 0$ so that A is in fact indefinite.[5] The indefiniteness of A is best seen by its eigenvalues, which are 0, 1 and -1, which obviously do not satisfy the condition of any definite matrix. On the other hand, the permutations of A can also reveal its indefiniteness. Take $\pi_4 = \{2, 3, 1\}$. Then $|A_1^{\pi_4}| = 1$,

$$|A_2^{\pi_4}| = \begin{vmatrix} 1 & 0 \\ 0 & -1 \end{vmatrix} = -1,$$

which fails the tests in Theorems 4.5 and 4.7.

4.10 Exercises

1. Verify that the sets and operations in Examples 4.1 to 4.5 satisfy the eight axioms of a vector space.
2. Show that the zero vector in a vector space forms a subspace.
3. Show that all the cases in Example 4.8 are subspaces.
4. Let $\mathbb{M}^{n \times n}$ be the vector space of all $n \times n$ square matrices. A symmetric matrix A is a square matrix equal to its transpose, that is, $A = A^T$. In other words, its elements $a_{ij} = a_{ji}$ for all $i, j = 1, \ldots, n$. Prove that the set of all $n \times n$ symmetric matrices is a subspace of $\mathbb{M}^{n \times n}$.
5. Show that $S = \{(x, 0) : x \in \mathbb{R}\}$ and $T = \{(0, y) : y \in \mathbb{R}\}$ are subspaces of $\mathbb{R}^2$.
   (a) Is $S \cup T$ a subspace of $\mathbb{R}^2$?
   (b) Is $S + T$ a subspace of $\mathbb{R}^2$?
6. Suppose that A is a nonempty subset of a vector space V. Show that the linear span of A is the intersection of all subspaces of V which contain A.
7. Determine whether each of the following sets is a subspace of the vector space $\mathbb{R}^3$. Justify your answers.
   (a) $B = \{(x_1, x_2, x_3) : x_1^2 + x_2^2 + x_3^2 < 1\}$,
   (b) $\mathbb{R}^2$,
   (c) $\mathbb{R}^3_+$,
   (d) a straight line passing through the origin.
8. Suppose that S and T are subspaces of a vector space V. Show that $S \cap T$ is a subspace of V.
9. Prove or disprove: Suppose that S and T are subspaces of a vector space V. Then $S \cup T$ is a subspace of V.
10. Prove or disprove: Suppose that S and T are subspaces of a vector space V. Then $S + T$ is a subspace of V.

---

[5] The vector e_i, $i = 1, \ldots, n$ is the i-th standard basis of $\mathbb{R}^n$. Therefore $e_2 = (0, 1, 0)$ in this example.

11. Consider the vector space $\mathbb{R}^3$.
    (a) Show that the set of vectors

$$B = \{(2, 1, 0), (0, 2, 1), (0, 0, 1)\}$$

    forms a basis for $\mathbb{R}^3$.
    (b) Let $\mathbf{x} = (1, 2, 3)$. Find the coordinates of $\mathbf{x}$ relative to the basis B.
12. Suppose that $V = \mathbb{M}^{2 \times 2}$ is the vector space of all 2×2 square matrices. Let

$$A = \begin{pmatrix} 1 & 2 \\ 3 & 4 \end{pmatrix}.$$

    What is the coordinates of A relative to the standard basis of $\mathbb{M}^{2 \times 2}$?
13. Let $f : V \to W$ be a linear transformation.
    (a) Prove that $f(\mathbf{0}) = \mathbf{0}$.
    (b) Show that the kernel of f is a subspace of V.
    (c) Show that the range of f, $f(V)$, is a subspace of W.
14. Suppose that the linear operator f on the vector space $\mathbb{R}^3$ is defined by

$$f(x, y, z) = (z, 0, y).$$

    (a) Find the kernel of f.
    (b) Use the Dimension Theorem to calculate the rank of f.
    (c) Find the matrix representation of f relative to the standard basis.
15. Let $f : V \to W$ be a linear transformation. Prove or disprove: The graph of f is a subspace of the product vector space $V \times W$.
16. Let V, W and Z be vector spaces and let $f : V \to W$ and $g : W \to Z$ be linear transformations. Show that $g \circ f$ is also a linear transformation.
17. Suppose that a linear operator f on $\mathbb{R}^3$ is represented by the matrix

$$A = \begin{pmatrix} 0 & 0 & 0 \\ 0 & 1 & 0 \\ 0 & 0 & -1 \end{pmatrix},$$

    relative to the standard basis.
    (a) Identify the kernel of f.
    (b) What is the nullity of f?
    (c) What is the rank of f?
    (d) Do your results above conform with the Dimension Theorem? Explain.
18. Let the linear transformations $f : \mathbb{R}^3 \to \mathbb{R}^3$ and $g : \mathbb{R}^3 \to \mathbb{R}^2$ be represented by the matrices

$$\begin{pmatrix} 3 & -2 & 0 \\ 1 & -1 & 4 \\ 5 & 5 & 3 \end{pmatrix} \quad \text{and} \quad \begin{pmatrix} 2 & -3 & 4 \\ 5 & 1 & 2 \end{pmatrix}$$

respectively. Find the matrix representation of the composite transformation $g \circ f$.

19. Show that dim $L(V, W) = \dim V \times \dim W$.
20. Show that each of the examples in Sect. 4.3.3 is a linear functional.
21. Verify that each of the examples in Sect. 4.4 is an inner product space.
22. Define a function $f : \mathbb{R}^2 \to \mathbb{R}$ by

$$f(x, y) = |x| + |y|.$$

Is $(\mathbb{R}, f)$ an inner product space? Explain.
23. Show that for every $\mathbf{x} \in V$, $\mathbf{0}^T \mathbf{x} = 0$.
24. Let V be a normed vector space. Show that (V, ρ) is a metric space, where for every $\mathbf{x}, \mathbf{y} \in V$,

$$\rho(\mathbf{x}, \mathbf{y}) = \|\mathbf{x} - \mathbf{y}\|.$$

25. Let V be a normed vector space. Prove the parallelogram equality: For every $\mathbf{x}, \mathbf{y} \in V$,

$$\|\mathbf{x} + \mathbf{y}\|^2 + \|\mathbf{x} - \mathbf{y}\|^2 = 2(\|\mathbf{x}\|^2 + \|\mathbf{y}\|^2).$$

26. Consider the relation $\perp$ on a vector space V. Determine if the relation is
    (a) reflexive,
    (b) transitive,
    (c) circular,
    (d) symmetric,
    (e) asymmetric,
    (f) antisymmetric.
27. Suppose that $(2\mathbf{x} + \mathbf{y}) \perp (\mathbf{x} - 3\mathbf{y})$ and $\|\mathbf{x}\| = 2\|\mathbf{y}\|$. Find the angle between $\mathbf{x}$ and $\mathbf{y}$.
28. Suppose that S is a one-dimensional subspace of a vector space V. Let $\{\mathbf{b}\}$ be a basis of S.
    (a) Define $S^\perp$, the orthogonal complement of S.
    (b) Show that $S^\perp$ is a subspace of V.
29. Let S be a subspace of a vector space V. Show that $S^\perp$ is also a subspace of V.
30. Give a geometric interpretation of the dot product in $\mathbb{R}^3$. What if one of the vector is a unit vector?
31. Suppose that $L(V)$ is the set of all linear operators on an n-dimensional vector space V. Let $D : L(V) \to \mathbb{R}$ be the determinant function. Prove or disprove: D is a linear functional.
32. Prove statements 2, 3, 5 and 6 in Sect. 4.6.
33. Let A, B and C be the matrix representations of three linear operators on a vector space. The three matrices are identical except for their first rows, which are $\mathbf{a}_1$, $\mathbf{b}_1$ and $\alpha \mathbf{a}_1 + \beta \mathbf{b}_1$ respectively, where $\alpha, \beta \in \mathbb{R}$. Show that

$$|C| = \alpha |A| + \beta |B|.$$

34. Show that if $BA = AC = I$, then $B = C$.
35. Let the matrix representation relative to the standard basis of the linear operator $f : \mathbb{R}^2 \to \mathbb{R}^2$ be

$$A = \begin{pmatrix} 2 & 3 \\ 3 & 4 \end{pmatrix}.$$

(a) Show that f is invertible.
(b) Show that f^{-1} is represented by the matrix

$$A^{-1} = \begin{pmatrix} -4 & 3 \\ 3 & -2 \end{pmatrix}.$$

(c) Is it true that $|A| = 1/|A^{-1}|$?
36. Let

$$A = \begin{pmatrix} 2 & 0 & -1 \\ 5 & 1 & 0 \\ 0 & 1 & 3 \end{pmatrix}, \quad B = \begin{pmatrix} 3 & -1 & 1 \\ -15 & 6 & -5 \\ 5 & -2 & 2 \end{pmatrix}.$$

Show that A is the inverse of B.
37. Find the inverses of the following matrices:

(a) $\begin{pmatrix} \cos\theta & \sin\theta \\ -\sin\theta & \cos\theta \end{pmatrix}$

(b) $\begin{pmatrix} -3 & 6 & -11 \\ 3 & -4 & 6 \\ 4 & -8 & 13 \end{pmatrix}$

38. Suppose that f and g are two invertible linear operators on a vector space V. Prove that

$$(g \circ f)^{-1}(\mathbf{x}) = (f^{-1} \circ g^{-1})(\mathbf{x})$$

for all $\mathbf{x} \in V$.
39. Find the inverse of

$$B = \begin{pmatrix} 1 & 1 & 0 \\ 0 & 1 & 0 \\ 0 & 0 & 1 \end{pmatrix}.$$

If B is the matrix representation of a linear operator f, find
(a) the kernel of f,
(b) the rank of f.

40. Consider the following system of linear equations:

$$x_1 - 2x_2 = -8,$$
$$5x_1 + 3x_2 = -1.$$

(a) Express the system in matrix form.
(b) Show that the solution exists.
(c) Solve x_1 and x_2 by a matrix inversion.

41. Given the system of linear equation

$$2x_1 - x_2 + 2x_3 = 2$$
$$x_1 + 10x_2 - 3x_3 = 5$$
$$-x_1 + x_2 + x_3 = -3.$$

(a) Express the system in matrix form.
(b) Find the solution for x_3.

42. Prove that if A and B are invertible, then $(AB)^{-1} = B^{-1}A^{-1}$.

43. Prove that if A is invertible, then $(A^{-1})^{-1} = A$.

44. Find the eigenvalues and normalized eigenvectors of the following matrices:

(i) $A = \begin{pmatrix} 3 & 4 \\ 4 & -3 \end{pmatrix}$, (ii) $A = \begin{pmatrix} 1 & 1 \\ 4 & 1 \end{pmatrix}$, (iii) $A = \begin{pmatrix} -5 & 2 \\ 2 & -2 \end{pmatrix}$, (iv) $A = \begin{pmatrix} 1 & 2 \\ 2 & 1 \end{pmatrix}$.

Show that $\sum_{i=1}^{n} \lambda_i = \sum_{i=1}^{n} a_{ii}$, $P^{\mathrm{T}} P = I$, $A = P \Lambda P^{\mathrm{T}}$ and $|\Lambda| = |A|$.

45. Let the matrix representation of a linear operator be

$$A = \begin{pmatrix} 3 & 6 \\ 1 & 4 \end{pmatrix}.$$

(a) Find the eigenvalues of A.
(b) Find the normalized eigenvectors of A.
(c) Are the eigenvectors orthogonal to each other? Why or why not?

46. Suppose that the linear operator f on $\mathbb{R}^3$ is represented by the matrix

$$\begin{pmatrix} 1 & 0 & 0 \\ 0 & 3 & 2 \\ 0 & 2 & 0 \end{pmatrix}$$

relative to the standard basis.
(a) Find the eigenvalues and the normalized eigenvectors.
(b) Determine the definiteness of the quadratic form.

47. Prove statement 1 in Sect. 4.8.2. Hint: Suppose that A has a complex eigenvalue $\lambda + i\beta$ where $i = \sqrt{-1}$ with the corresponding eigenvector $\mathbf{x} + i\mathbf{y}$. Then

$$A(\mathbf{x} + i\mathbf{y}) = (\lambda + i\beta)(\mathbf{x} + i\mathbf{y}).$$

Show that $\beta = 0$.

48. Prove that for distinct eigenvalues λ_1 and λ_2 of a symmetric operator A, the corresponding eigenvectors $\mathbf{x}_1$ and $\mathbf{x}_2$ are orthogonal.

49. Prove the Spectral Theorem in Sect. 4.8.2. Hint: Recall that columns in P are pairwise orthonormal, that is,

$$\mathbf{x}_i^{\mathrm{T}}\mathbf{x}_j = \begin{cases} 1, & \text{if } i = j, \\ 0, & \text{otherwise.} \end{cases}$$

50. Prove statement 6 in Sect. 4.8.2. Hint: first show that $|P| = \pm 1$ and use the Spectral Theorem.

51. Prove statements 8, 9 and 10 in Sect. 4.8.2.

52. Let X be an $n \times k$ matrix where $k \leq n$ and the columns of X are linearly independent so that $(X^{\mathrm{T}}X)^{-1}$ exists.
    (a) Show that

    $$M_1 = X(X^{\mathrm{T}}X)^{-1}X^{\mathrm{T}}$$

    and

    $$M_2 = I_n - X(X^{\mathrm{T}}X)^{-1}X^{\mathrm{T}}$$

    are idempotent matrices.
    (b) Suppose that the diagonal decomposition of $M_1 = P\Lambda_1 P^{\mathrm{T}}$. Show that $M_2 = P\Lambda_2 P^{\mathrm{T}}$ where $\Lambda_1 + \Lambda_2 = I_n$.

53. Let A be an $n \times n$ matrix (not necessary symmetric) which has nonnegative elements and satisfies $\sum_{j=1}^{n} a_{ij} = 1$ for $i = 1, \ldots, n$. Show that A^{T} has at least one unit eigenvalue.
    Hint: Consider the simplex

    $$S = \left\{ \mathbf{x} \in \mathbb{R}^n : x_i \geq 0, \sum_{i=1}^{n} x_i = 1 \right\}.$$

    Show that the linear function f represented by A^{T} is a linear operator on S, that is, for every $\mathbf{x} \in S$, $f(\mathbf{x})$ is also in S. Then show that $f : S \to S$ satisfies the hypothesis of the Brouwer Fixed Point Theorem. A is called a transition matrix and is important in the study of Markov process in macroeconomics (see Ljungqvist and Sargent, 2004, 30–31).

54. Let

$$A = \begin{pmatrix} 0 & 0 & 0 \\ 0 & -3 & 0 \\ 0 & 0 & 2 \end{pmatrix}, \quad B = \begin{pmatrix} 1 & 0 & 0 \\ 0 & 3 & 0 \\ 0 & 0 & 0 \end{pmatrix}, \quad C = \begin{pmatrix} -3 & 0 & 0 \\ 0 & -1 & 0 \\ 0 & 0 & -2 \end{pmatrix}.$$

(a) Determine the definiteness of A, B and C.
(b) Determine the definiteness of $A + B + C$.

55. Let X be an $n \times k$ matrix where $k \leq n$ and the columns of X are linearly independent. Show that $X^T X$ is positive definite.

56. Let A be a positive definite matrix. Show that $A = QQ^T$ where Q is an invertible matrix.

57. Let A be a positive definite matrix. Prove that A^{-1} is also positive definite.

58. Suppose that f is a symmetric linear operator on a n-dimensional vector space V. Prove the **polarization identity**: For every $\mathbf{x}, \mathbf{y} \in V$,

$$\mathbf{x}^T f(\mathbf{y}) = \frac{1}{4} Q(\mathbf{x} + \mathbf{y}) - \frac{1}{4} Q(\mathbf{x} - \mathbf{y}).$$

References

Hoffman, K., & Kunze, R. (1971). *Linear algebra*, Second Edition. Englewood Cliffs: Prentice-Hall.

Ljungqvist, L., & Thomas, J. S. (2004). *Recursive macroeconomic theory*, Second Edition. Cambridge: The MIT Press.

Vector Calculus

5

In this chapter we introduce multivariate differential calculus, which is an important tool in economic modelling. Concepts developed in the previous chapters are applied to the Euclidean space, which is both a metric space and a normed vector space.

5.1 Vector Differentiation

5.1.1 Introduction

Recall that in single variable calculus, the derivative of a function $f : \mathbb{R} \to \mathbb{R}$ at a point x is defined by

$$f'(x) = \lim_{h \to 0} \frac{f(x+h) - f(x)}{h},$$

provided that the limit exists. The limit can be expressed as

$$\lim_{h \to 0} \frac{f(x+h) - f(x) - f'(x)h}{h} = 0. \tag{5.1}$$

If the limit exists at every point in a set $E \subseteq \mathbb{R}$, we say that f is differentiable on E. Differentiability implies that the function is continuous on E, but the converse is not true. There are many examples of continuous functions which are not differentiable.

Example 5.1 Let $f(x) = x^2$. The limit in (5.1) is

$$\lim_{h \to 0} \frac{(x+h)^2 - x^2 - f'(x)h}{h} = \lim_{h \to 0} \frac{x^2 + 2hx + h^2 - x^2 - f'(x)h}{h}$$

© Springer Nature Switzerland AG 2019
K. Yu, *Mathematical Economics*, Springer Texts in Business and Economics,
https://doi.org/10.1007/978-3-030-27289-0_5

$$= \lim_{h \to 0} \frac{2hx + h^2 - f'(x)h}{h}$$

$$= \lim_{h \to 0} (2x + h - f'(x))$$

$$= 2x - f'(x),$$

which gives $f'(x) = 2x$. For any particular point, say $x = -1$, the derivative $f'(-1) = -2$. The last term in the numerator of the limit in (5.1) is $f'(x)h = -2h$, which is a linear function in h. Geometrically, $f'(x)$ is the slope of the function $f(x)$ at the point x.

Now we extend the concept of differentiation to the case of functions of several variables. Recall from Chaps. 3 and 4 that the Euclidean space $\mathbb{R}^n$ is both a metric space and a normed vector space, with the metric defined as

$$\rho(\mathbf{x}, \mathbf{y}) = \|\mathbf{x} - \mathbf{y}\|,$$

for every $\mathbf{x}, \mathbf{y} \in \mathbb{R}^n$. Suppose that S is an open set in $\mathbb{R}^n$, f maps S into $\mathbb{R}^m$, and $\mathbf{x} \in S$. If there exists a linear transformation, $Df(\mathbf{x}) : \mathbb{R}^n \to \mathbb{R}^m$ such that, for $\mathbf{h} \in \mathbb{R}^n$,

$$\lim_{\mathbf{h} \to 0} \frac{\|f(\mathbf{x} + \mathbf{h}) - f(\mathbf{x}) - Df(\mathbf{x})(\mathbf{h})\|}{\|\mathbf{h}\|} = 0, \tag{5.2}$$

then f is differentiable at $\mathbf{x}$ and $Df(\mathbf{x})$ is called the **derivative** of f at $\mathbf{x}$. If f is differentiable at every point in S, we say that f is differentiable on S. Since $Df(\mathbf{x})$ is linear, it can be represented by a $m \times n$ matrix relative to the standard bases in $\mathbb{R}^m$ and $\mathbb{R}^n$. If $m = n$, $Df(\mathbf{x})$ becomes a linear operator. In this case, the determinant of $Df(\mathbf{x})$ is called the **Jacobian**, $J_f(\mathbf{x})$, of f at $\mathbf{x}$.

Example 5.2 Suppose that $f : \mathbb{R}^n \to \mathbb{R}^m$ is a linear function. Then f can be represented by a $m \times n$ matrix A relative to the standard bases such that $f(\mathbf{x}) = A\mathbf{x}$. The limit in definition (5.2) becomes

$$\lim_{\mathbf{h} \to 0} \frac{\|A(\mathbf{x} + \mathbf{h}) - A\mathbf{x} - Df(\mathbf{x})(\mathbf{h})\|}{\|\mathbf{h}\|} = \lim_{\mathbf{h} \to 0} \frac{\|A\mathbf{x} + A\mathbf{h} - A\mathbf{x} - Df(\mathbf{x})(\mathbf{h})\|}{\|\mathbf{h}\|}$$

$$= \lim_{\mathbf{h} \to 0} \frac{\|A\mathbf{h} - Df(\mathbf{x})(\mathbf{h})\|}{\|\mathbf{h}\|}.$$

It is clear that the limit is zero if $Df(\mathbf{x}) = A$ for all $\mathbf{x} \in \mathbb{R}^n$. If $m = n$, then $Df(\mathbf{x})$ is a square matrix and the Jacobian is $J_f(\mathbf{x}) = |A|$, the determinant of A. For $m = n = 1$, the function reduces to the single variable case that $f(x) = ax$ so that $f'(x) = a$.

Some important properties of $Df(\mathbf{x})$ are listed below:

1. If f is differentiable, then $Df(\mathbf{x})$ is unique.[1]
2. $f(\mathbf{x}) + Df(\mathbf{x})\mathbf{h}$ is a linear approximation of $f(\mathbf{x} + \mathbf{h})$.
3. A function f is continuous if it is differentiable.
4. Let $F(\mathbb{R}^n, \mathbb{R}^m)$ be the collection of all differentiable functions from $\mathbb{R}^n$ to $\mathbb{R}^m$. The derivatives are linear functions on $F(\mathbb{R}^n, \mathbb{R}^m)$, that is, if $f, g \in F(\mathbb{R}^n, \mathbb{R}^m)$, then for any real number α,

$$D(\alpha f + g)(\mathbf{x}) = \alpha Df(\mathbf{x}) + Dg(\mathbf{x}).$$

5. Chain rule: Suppose that $f : \mathbb{R}^n \to \mathbb{R}^m$ and $g : \mathbb{R}^m \to \mathbb{R}^l$ are differentiable. Then $D(g \circ f)(\mathbf{x})$ is a linear function from $\mathbb{R}^n$ into $\mathbb{R}^l$ such that

$$D(g \circ f)(\mathbf{x}) = Dg(f(\mathbf{x}))Df(\mathbf{x}).$$

6. Product rule: Suppose that f and g are functionals on S and define $fg(\mathbf{x}) = f(\mathbf{x})g(\mathbf{x})$. Then

$$D(fg)(\mathbf{x}) = f(\mathbf{x})Dg(\mathbf{x}) + g(\mathbf{x})Df(\mathbf{x}).$$

Next we define the partial derivatives of f. Suppose that f maps an open set $S \subseteq \mathbb{R}^n$ into $\mathbb{R}^m$. Let $(\mathbf{e}_1, \mathbf{e}_2, \ldots, \mathbf{e}_n)$ and $(\mathbf{u}_1, \mathbf{u}_2, \ldots, \mathbf{u}_m)$ be the standard bases for $\mathbb{R}^n$ and $\mathbb{R}^m$ respectively. Then f can be expressed as its component functionals relative to $(\mathbf{u}_1, \mathbf{u}_2, \ldots, \mathbf{u}_m)$:

$$f(\mathbf{x}) = \begin{pmatrix} f_1(\mathbf{x}) \\ f_2(\mathbf{x}) \\ \vdots \\ f_m(\mathbf{x}) \end{pmatrix},$$

where $f_i(\mathbf{x}) = f(\mathbf{x})^{\mathsf{T}}\mathbf{u}_i$, for $i = 1, \ldots, m$. The partial derivative of f_i with respect to x_j is defined as

$$\frac{\partial f_i(\mathbf{x})}{\partial x_j} = \lim_{t \to 0} \frac{f_i(\mathbf{x} + t\mathbf{e}_j) - f_i(\mathbf{x})}{t},$$

provided that the limit exists. The following theorem reveals the matrix representation of the linear function $Df(\mathbf{x})$ relative to the standard bases.

---

[1] See Rudin (1976, p. 213) for a proof.

Theorem 5.1 *Suppose that f defined above is differentiable at a point $\mathbf{x} \in S$. Then the partial derivatives exist and*

$$Df(\mathbf{x})\mathbf{e}_j = \sum_{i=1}^{m} \frac{\partial f_i(\mathbf{x})}{\partial x_j} \mathbf{u}_i, \quad j = 1, \ldots, n.$$

See Rudin (1976, p. 215–6) for a proof of this theorem. Consequently, the matrix representation of $Df(\mathbf{x})$ relative to the above standard bases is

$$\begin{pmatrix} \partial f_1(\mathbf{x})/\partial x_1 & \partial f_1(\mathbf{x})/\partial x_2 & \cdots & \partial f_1(\mathbf{x})/\partial x_n \\ \partial f_2(\mathbf{x})/\partial x_1 & \partial f_2(\mathbf{x})/\partial x_2 & \cdots & \partial f_2(\mathbf{x})/\partial x_n \\ \vdots & \vdots & & \vdots \\ \partial f_m(\mathbf{x})/\partial x_1 & \partial f_m(\mathbf{x})/\partial x_2 & \cdots & \partial f_m(\mathbf{x})/\partial x_n \end{pmatrix}.$$

5.1.2   Functional Analysis

In the case of $m = 1$, $Df(\mathbf{x})$ is a linear functional and is often called the **gradient** of f at $\mathbf{x}$ and is denoted by $\nabla f(\mathbf{x})$. Although $Df(\mathbf{x})$ is a $1 \times n$ matrix, we usually write $\nabla f(\mathbf{x})$ as a column vector like $\mathbf{x}$. Therefore in partial derivative form we have

$$\nabla f(\mathbf{x}) = \begin{pmatrix} \partial f(\mathbf{x})/\partial x_1 \\ \partial f(\mathbf{x})/\partial x_2 \\ \vdots \\ \partial f(\mathbf{x})/\partial x_n \end{pmatrix}.$$

Example 5.3 Suppose $f(x_1, x_2) = a x_1^{\alpha} x_2^{\beta}$, then

$$\nabla f(\mathbf{x}) = \begin{pmatrix} \partial f(\mathbf{x})/\partial x_1 \\ \partial f(\mathbf{x})/\partial x_2 \end{pmatrix} = \begin{pmatrix} \alpha a x_1^{\alpha-1} x_2^{\beta} \\ \beta a x_1^{\alpha} x_2^{\beta-1} \end{pmatrix}.$$

Notice that $\nabla f(\mathbf{x})$ is a linear functional for any particular value of $\mathbf{x}$. For example, when $\mathbf{x} = (1, 1)$,

$$\nabla f(1, 1) = \begin{pmatrix} \alpha a (1)^{\alpha-1} (1)^{\beta} \\ \beta a (1)^{\alpha} (1)^{\beta-1} \end{pmatrix} = \begin{pmatrix} \alpha a \\ \beta a \end{pmatrix}.$$

In the above example, a different $\mathbf{x}$ of course gives a different functional. We can, however, view the gradient as a (non)linear function which maps vectors in S into vectors in $\mathbb{R}^n$. In other words, we can treat the matrix representation of the gradients as vectors in $\mathbb{R}^n$. Then we can replace f with $\nabla f(\mathbf{x})$ in (5.2) and the resulting linear operator, if it exists, is called the second derivative or **Hessian** of f at $\mathbf{x}$, $\nabla^2 f(\mathbf{x})$.

We say that f is twice differentiable at $\mathbf{x}$. The matrix representation of the Hessian is by definition a $n \times n$ square matrix in terms of the partial derivatives, that is,

$$\nabla^2 f(\mathbf{x}) = \begin{pmatrix} \partial^2 f(\mathbf{x})/\partial x_1 \partial x_1 & \partial^2 f(\mathbf{x})/\partial x_1 \partial x_2 & \cdots & \partial^2 f(\mathbf{x})/\partial x_1 \partial x_n \\ \partial^2 f(\mathbf{x})/\partial x_2 \partial x_1 & \partial^2 f(\mathbf{x})/\partial x_2 \partial x_2 & \cdots & \partial^2 f(\mathbf{x})/\partial x_2 \partial x_n \\ \vdots & \vdots & & \vdots \\ \partial^2 f(\mathbf{x})/\partial x_n \partial x_1 & \partial^2 f(\mathbf{x})/\partial x_n \partial x_2 & \cdots & \partial^2 f(\mathbf{x})/\partial x_n \partial x_n \end{pmatrix}.$$

If f is twice differentiable at $\mathbf{x}$ and each element of the Hessian is a continuous function at $\mathbf{x}$, f is called a $\mathcal{C}^2$ function. Also, the Schwarz Identity (often called Young's Theorem) states that the Hessian is symmetric, that is, for $i \neq j$,

$$\frac{\partial^2 f(\mathbf{x})}{\partial x_i \partial x_j} = \frac{\partial^2 f(\mathbf{x})}{\partial x_j \partial x_i}.$$

Table 5.1 lists the gradients and Hessians of some commonly used functionals. The square matrix $\mathbf{yy}^\mathsf{T}$ in example 4 is called the **outer product** of the vector $\mathbf{y}$ with itself. In example 5, if A is symmetric, the gradient and Hessian of $\mathbf{x}^\mathsf{T} A \mathbf{x}$ are $2A\mathbf{x}$ and $2A$ respectively. Also, if A is the identity matrix, the results reduce to those in example 3.

Geometrically, the partial derivative of a functional f with respect to the standard basis, $\partial f(\mathbf{x})/\partial x_j$, is the rate of increase of $f(\mathbf{x})$ at $\mathbf{x}$ in the direction of the j-th coordinate. The definition can be modified to find the rate of increase in an arbitrary direction. Suppose $\mathbf{u} \in \mathbb{R}^n$ is a unit vector, with $\|\mathbf{u}\| = 1$. Then the **directional derivative** of f at $\mathbf{x}$ in the direction of $\mathbf{u}$ is defined as

$$D_\mathbf{u} f(\mathbf{x}) = \lim_{t \to 0} \frac{f(\mathbf{x} + t\mathbf{u}) - f(\mathbf{x})}{t}.$$

It can be shown that if f is differentiable,[2]

$$D_\mathbf{u} f(\mathbf{x}) = \nabla f(\mathbf{x})^\mathsf{T} \mathbf{u}.$$

Table 5.1 Some common functionals

Example	$f(\mathbf{x})$	$\nabla f(\mathbf{x})$	$\nabla^2 f(\mathbf{x})$
1	$\mathbf{y}^\mathsf{T} \mathbf{x}$	$\mathbf{y}$	$0_{n \times n}$
2	$\mathbf{x}^\mathsf{T} A \mathbf{y}$	$A\mathbf{y}$	$0_{n \times n}$
3	$\mathbf{x}^\mathsf{T} \mathbf{x}$	$2\mathbf{x}$	$2I_n$
4	$\mathbf{y}^\mathsf{T} \mathbf{xx}^\mathsf{T} \mathbf{y}$	$2\mathbf{yy}^\mathsf{T} \mathbf{x}$	$2\mathbf{yy}^\mathsf{T}$
5	$\mathbf{x}^\mathsf{T} A \mathbf{x}$	$(A + A^\mathsf{T})\mathbf{x}$	$A + A^\mathsf{T}$

---

[2] See Exercise 6 at the end of the chapter.

Fig. 5.1 Gradient of a
functional

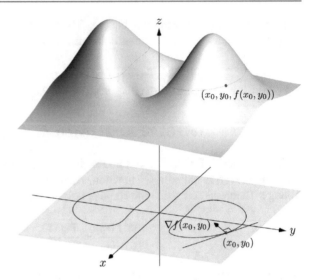

It is clear that $D_{\mathbf{u}}f(\mathbf{x})$ achieves its maximum value when $\mathbf{u}$ has the same direction as $\nabla f(\mathbf{x})$. In other words, the gradient points to the direction in which the value of f is increasing at the fastest rate, with the rate of increase given by $\|\nabla f(\mathbf{x})\|$. Figure 5.1 illustrates the relationship between the gradient and the graph of f in the case of $n = 2$. Notice that $\nabla f(x_0, y_0)$ is orthogonal to the tangent line to the level curve at (x_0, y_0).

Theorem 5.2 (Mean Value Theorem) *Suppose that f is a differentiable functional on a convex set $S \subseteq \mathbb{R}^n$. For all $\mathbf{x}, \mathbf{y} \in S$, there exists a point $\mathbf{z} = \alpha\mathbf{x} + (1-\alpha)\mathbf{y}$ for some $\alpha \in [0, 1]$ such that*

$$f(\mathbf{x}) - f(\mathbf{y}) = \nabla f(\mathbf{z})^{\mathrm{T}}(\mathbf{x} - \mathbf{y}).$$

5.2 Approximation of Functions

Unlike, say, Newton's law of gravity, the exact functional forms in economic modelling are rarely known. Often we choose a functional form to satisfy certain structural restrictions from the model. For analytical convenience, we also want to have functional forms that are easy to manipulate. In this section we look at the approximation of functions. The following theorem provides the starting point. It assures us that any continuous real valued function defined on a closed interval can be approximated uniformly by polynomials.

Theorem 5.3 (Stone-Weierstrass Theorem) *Let $f : [a, b] \to \mathbb{R}$ be a continuous function. Then there exists a sequence of polynomials $P_n(x)$ that converges uniformly to f on $[a, b]$.*

The proof of this remarkable result can be found in Brosowski and Deutsch (1981). If the function f is differentiable, the polynomial approximation becomes the well-known test Taylor series.

5.2.1 Taylor Series

Let $f(x)$ be a single variable function of x which is differentiable to any order s. The Taylor expansion of f centred at a point x^* is given by

$$f(x) = \sum_{s=0}^{\infty} \frac{f^{(s)}(x^*)}{s!}(x - x^*)^s$$

$$= f(x^*) + f'(x^*)(x - x^*) + \frac{1}{2}f''(x^*)(x - x^*)^2 + \frac{1}{3!}f'''(x^*)(x - x^*)^3 + \cdots$$

Figure 5.2 depicts approximations of $y = \sin x$ at $x^* = 0$. Notice that the approximation becomes better as the order increases. Nevertheless, the approximated values deviate dramatically from the true values once x exceed a certain distance from x^*.

In economic applications we often take the approximation up to the first or second derivative term, which we called first-order and second-order approximations respectively.

In the multivariate cases Taylor approximations normally go up to second order only because of the complexity of higher derivatives.

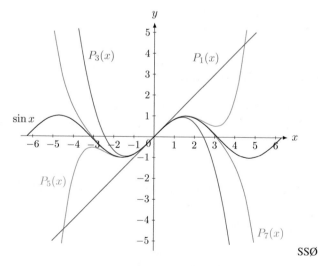

Fig. 5.2 Taylor approximations of $y = \sin x$

Theorem 5.4 (Taylor's Theorem) *Suppose f is a C^2 functional on a convex set $S \subseteq \mathbb{R}^n$. Then for $\mathbf{x}_0, \mathbf{x} \in S$,*

$$f(\mathbf{x}) = f(\mathbf{x}_0) + \nabla f(\mathbf{x}_0)^{\mathrm{T}}(\mathbf{x} - \mathbf{x}_0) + \frac{1}{2}(\mathbf{x} - \mathbf{x}_0)^{\mathrm{T}} \nabla^2 f(\mathbf{x}_0)(\mathbf{x} - \mathbf{x}_0) + r(\mathbf{x}_0, \mathbf{x}),$$

where $\lim_{\mathbf{x} \to \mathbf{x}_0} r(\mathbf{x}_0, \mathbf{x}) = 0$.

Taylor's theorem provides us a second-order approximation (sometimes called a 2-jet) of an arbitrary C^2 functional about a point $\mathbf{x}_0$ by ignoring the remainder term $r(\mathbf{x}_0, \mathbf{x})$. In dynamic models $\mathbf{x}_0$ is often chosen as the steady-state (long-run) value of $\mathbf{x}$.

A functional which can provide a second-order approximation to an arbitrary C^2 functional is called a **flexible functional form** and is popular among applied econometricians. The quadratic functional form is defined as

$$f(\mathbf{x}) = \alpha_0 + \boldsymbol{\alpha}^{\mathrm{T}}\mathbf{x} + \frac{1}{2}\mathbf{x}^{\mathrm{T}}A\mathbf{x}, \tag{5.3}$$

where $\alpha_0 \in \mathbb{R}, \boldsymbol{\alpha} \in \mathbb{R}^n$ and $A = A^{\mathrm{T}}$, an $n \times n$ symmetric matrix, are unknown parameters to be estimated from the regression model. The total number of unknown parameters is $1 + n + n(n+1)/2$ and so if n is big the regression needs a large number of observations. Another popular flexible functional form is the translog function, defined as

$$\log f(\mathbf{x}) = \beta_0 + \boldsymbol{\beta}^{\mathrm{T}} \log \mathbf{x} + \frac{1}{2} \log \mathbf{x}^{\mathrm{T}} B \log \mathbf{x},$$

where $\log \mathbf{x} = (\log x_1 \ \log x_2 \cdots \ \log x_n)^{\mathrm{T}}$, and $\beta_0 \in \mathbb{R}, \boldsymbol{\beta} \in \mathbb{R}^n$ and $B = B^{\mathrm{T}}$, an $n \times n$ symmetric matrix, are again the unknown parameters.

5.2.2 Logarithmic Approximation

The function $f(x) = \log(1+x)$ is tangent to the 45 degree line at $x = 0$ (see Fig. 5.3 and Exercise 12 below). Therefore for small values of x, a useful approximation is

$$\log(1 + x) \simeq x.$$

For example, let P_t be the price level in period t and let $p_t = \log P_t$. Then

$$\Delta p_t = p_t - p_{t-1} = \log P_t - \log P_{t-1}$$

$$= \log\left(\frac{P_t}{P_{t-1}}\right) = \log\left(1 + \frac{P_t - P_{t-1}}{P_{t-1}}\right)$$

$$= \log(1 + \pi_t) \simeq \pi_t. \tag{5.4}$$

Fig. 5.3 Logarithmic approximation

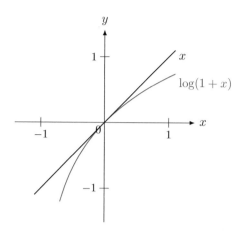

That is, the log difference of price levels between the two periods is approximately equal to the inflation rate π_t. Similarly, the log difference of output levels between two periods can be used as the growth rate in practice.

On the other hand, for x around 1, the approximation can be modified to become

$$\log x \simeq x - 1. \tag{5.5}$$

5.2.3 Log-Linear Approximation

Another useful approximation in economics is the log-linear approximation. Let $f(x)$ be any differentiable function. Using the Taylor first-order approximation at a point x^*, we have

$$\begin{aligned}
f(x) &\simeq f(x^*) + f'(x^*)(x - x^*) \\
&= f(x^*) + x^* f'(x^*)\left(\frac{x}{x^*} - 1\right) \\
&\simeq f(x^*) + x^* f'(x^*) \log\left(\frac{x}{x^*}\right) \quad \text{(using (5.5))} \\
&= f(x^*) + x^* f'(x^*)(\log x - \log x^*).
\end{aligned}$$

The above result is often written as

$$f(x) \simeq f(x^*) + x^* f'(x^*)\hat{x}$$

where $\hat{x} = \log x - \log x^*$. In view of (5.4), $\hat{x}$ can be interpreted as the proportion of x deviated from x^*. For example, let $f(x) = e^x$. Then the log-linear approximation at $x^* = 1$ is

Fig. 5.4 Log-linear
approximation

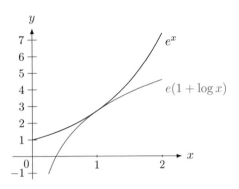

$$e^x \simeq e^1 + (1)e^1 (\log x - \log(1))$$
$$= e + e \log x$$
$$= e(1 + \log x).$$

Figure 5.4 depicts e^x and its log-linear approximation at $x^* = 1$. See Clausen (2006) for more discussions on the topic. See Wickens (2011, p. 506) for an application of the log-linear approximation in solving a real business cycle model.

5.3   Inverse Functions

Recall that if f is a linear operator on a vector space V and f is a bijection, then f is invertible. In matrix representation with respect to a basis of V, we have $\mathbf{y} = A\mathbf{x}$ and $\mathbf{x} = A^{-1}\mathbf{y}$. Now suppose that f is nonlinear and continuously differentiable ($\mathcal{C}^1$) function. Under what conditions can we be sure that f has an inverse in a neighbourhood? It may be very difficult to find an explicit inverse of f. But can we find instead the derivative of f^{-1}? The answers to both questions are the results of the inverse function theorem:

Theorem 5.5 (Inverse Function Theorem) *Suppose f is a $\mathcal{C}^1$ mapping of an open set $S \subseteq \mathbb{R}^n$ into $\mathbb{R}^n$. If the Jacobian $J_f(\mathbf{x}) \neq 0$ at some point $\mathbf{x} \in S$ and $\mathbf{y} = f(\mathbf{x})$. Then*

1. there exist open sets U and V in $\mathbb{R}^n$ such that $\mathbf{x} \in U$, $\mathbf{y} \in V$, $f : U \to V$ is a bijection;
2. the derivative of the inverse function $f^{-1} : V \to U$ is given by

$$Df^{-1}(\mathbf{y}) = [Df(\mathbf{x})]^{-1}. \tag{5.6}$$

In other words, the function f is bijective at each point $\mathbf{x} \in U$. An invertible function $f : \mathbb{R}^n \to \mathbb{R}^n$ is often called a **diffeomorphism**.

Example 5.4 Let $f : \mathbb{R}^2 \to \mathbb{R}^2$ be given by

$$f(\mathbf{x}) = ((x_1^2 + x_2^2)/2, \; x_1 x_2).$$

The range of f is $\{(y_1, y_2) \in \mathbb{R}^2 : y_1 \geq 0\}$. The derivative of f is

$$Df(\mathbf{x}) = \begin{pmatrix} x_1 & x_2 \\ x_2 & x_1 \end{pmatrix},$$

with Jacobian $J_f(\mathbf{x}) = x_1^2 - x_2^2$. Therefore f is not invertible when $x_1 = \pm x_2$. Put $\mathbf{x} = (1, 0)$ and let $\mathbf{y} = f(\mathbf{x})$. Then by (5.6)

$$Df^{-1}(\mathbf{y}) = [Df(\mathbf{x})]^{-1} = \begin{pmatrix} 1 & 0 \\ 0 & 1 \end{pmatrix}^{-1} = \begin{pmatrix} 1 & 0 \\ 0 & 1 \end{pmatrix}$$

Sometimes we have a system of n nonlinear equations with n endogenous variables $\mathbf{x}$ and m exogenous variables $\boldsymbol{\theta}$ (often called parameters). Instead of finding an explicit formula for $\mathbf{x}$ in terms of $\boldsymbol{\theta}$, we can study the effect of changes in $\mathbf{x}$ for changes in $\boldsymbol{\theta}$, which we call comparative statics. The implicit function theorem is a useful tool for doing that. If $f(\mathbf{x}, \boldsymbol{\theta})$ is a function of $\mathbf{x} \in \mathbb{R}^n$ and $\boldsymbol{\theta} \in \mathbb{R}^m$ into $\mathbb{R}^n$, then we denote the derivative of f with respect to $\mathbf{x}$ by $D_{\mathbf{x}} f(\mathbf{x}, \boldsymbol{\theta})$ and similarly $\boldsymbol{\theta}$ by $D_{\boldsymbol{\theta}} f(\mathbf{x}, \boldsymbol{\theta})$.

Theorem 5.6 (Implicit Function Theorem) *Let f be a C^1 mapping of an open set $S \subseteq \mathbb{R}^{n+m}$ into $\mathbb{R}^n$ such that $f(\mathbf{x}, \boldsymbol{\theta}) = \mathbf{0}$ for some point $(\mathbf{x}, \boldsymbol{\theta}) \in S$. If $D_{\mathbf{x}} f(\mathbf{x}, \boldsymbol{\theta})$ is invertible, then there exist open sets $U \subset \mathbb{R}^{n+m}$ and $W \subset \mathbb{R}^m$, with $(\mathbf{x}, \boldsymbol{\theta}) \in U$, $\boldsymbol{\theta} \in W$, and a C^1 mapping $g : W \to \mathbb{R}^n$ such that*

$$\mathbf{x} = g(\boldsymbol{\theta}), \tag{5.7}$$

$$f(g(\boldsymbol{\phi}), \boldsymbol{\phi}) = \mathbf{0} \quad \text{for all } \boldsymbol{\phi} \in W, \tag{5.8}$$

$$Dg(\boldsymbol{\theta}) = -[D_{\mathbf{x}} f(\mathbf{x}, \boldsymbol{\theta})]^{-1} D_{\boldsymbol{\theta}} f(\mathbf{x}, \boldsymbol{\theta}). \tag{5.9}$$

Equation (5.7) means that the solution of the endogenous variable $\mathbf{x}$ exists as a function of the exogenous variable $\boldsymbol{\theta}$ in the neighbourhood W. Equation (5.8) ensures that the solutions $g(\boldsymbol{\phi})$ for all $\boldsymbol{\phi} \in W$ satisfy the system of equations $f(\mathbf{x}, \boldsymbol{\theta}) = \mathbf{0}$. And finally, although an explicit solution of the function g may be difficult to obtain, we can get the derivative of g by Eq. (5.9). The elements of the matrix $Dg(\boldsymbol{\theta})$ are the rates of change of the endogenous variables in $\mathbf{x}$ with respect to the exogenous variables in $\boldsymbol{\theta}$.

Example 5.5 We look at the simple case that f is a linear transformation from $\mathbb{R}^{n+m}$ to $\mathbb{R}^n$. Then f can be represented by a matrix relative to the standard basis. We are interested in all the vectors $(\mathbf{x}, \boldsymbol{\theta})$ that belong to the kernel of f. That is,

$$f(\mathbf{x}, \boldsymbol{\theta}) = \begin{bmatrix} A & B \end{bmatrix} \begin{bmatrix} \mathbf{x} \\ \boldsymbol{\theta} \end{bmatrix} = \mathbf{0} \in \mathbb{R}^n,$$

where A is an $n \times n$ matrix and B is an $n \times m$ matrix. This means that

$$A\mathbf{x} + B\boldsymbol{\theta} = \mathbf{0}.$$

If A is invertible, then

$$\mathbf{x} = g(\boldsymbol{\theta}) = -A^{-1}B\boldsymbol{\theta},$$

which is an explicit solution in Eq. (5.7). For any vector $\boldsymbol{\phi} \in \mathbb{R}^m$,

$$f(g(\boldsymbol{\phi}), \boldsymbol{\phi}) = A(-A^{-1}B\boldsymbol{\phi}) + B\boldsymbol{\phi} = \mathbf{0},$$

which verifies Eq. (5.8). Finally, the rate of change of $\mathbf{x}$ with respect to $\boldsymbol{\theta}$ is

$$Dg(\boldsymbol{\theta}) = -A^{-1}B.$$

This is Eq. (5.9) with the facts that $D_{\mathbf{x}}f(\mathbf{x}, \boldsymbol{\theta}) = A$ and $D_{\boldsymbol{\theta}}f(\mathbf{x}, \boldsymbol{\theta}) = B$.

Example 5.6 Consider the following Keynesian model for the aggregate economy:

$$Y = C[(1-t)Y] + I(r, Y) + G,$$

$$M/P = L(Y, r),$$

The first equation is often called the IS equation and the second LM equation. In this model, Y (output) and r (interest rate) are endogenous and P (price), G (government expenditure), t (tax rate) and M (money supply) are exogenous. C, I and L are functions for consumption, investment and money demand respectively. In the context of Theorem 5.6, we have $n = 2, m = 4$, with $\mathbf{x} = [Y \ r]^{\mathrm{T}}$ and $\boldsymbol{\theta} = [t \ G \ M \ P]^{\mathrm{T}}$. The function f is

$$f(\mathbf{x}, \boldsymbol{\theta}) = \begin{pmatrix} Y - C[(1-t)Y] - I(r, Y) - G \\ M/P - L(Y, r) \end{pmatrix}.$$

The derivative of f with respect to $\mathbf{x}$ is

$$D_{\mathbf{x}}f(\mathbf{x}, \boldsymbol{\theta}) = \begin{pmatrix} 1 - C'[\cdot](1-t) - \partial I/\partial Y & -\partial I/\partial r \\ -\partial L/\partial Y & -\partial L/\partial r \end{pmatrix}. \tag{5.10}$$

In order for the model to work and to have an implicit solution, $D_{\mathbf{x}} f(\mathbf{x}, \boldsymbol{\theta})$ must be invertible, that is, the Jacobian cannot be zero. In this IS-LM model, we normally assume that $1 - C'[\cdot](1-t) - \partial I/\partial Y$ is a positive fraction, $\partial I/\partial r < 0$, $\partial L/\partial Y > 0$, and $\partial L/\partial r < 0$. Therefore the Jacobian is

$$J = -[1 - C'[\cdot](1-t) - \partial I/\partial Y](\partial L/\partial r) - (\partial I/\partial r)(\partial L/\partial Y) > 0.$$

The derivative of f with respect to $\boldsymbol{\theta}$ is

$$D_{\boldsymbol{\theta}} f(\mathbf{x}, \boldsymbol{\theta}) = \begin{pmatrix} C'[\cdot]Y & -1 & 0 & 0 \\ 0 & 0 & 1/P & -M/P^2 \end{pmatrix}. \tag{5.11}$$

Using (5.9) in Theorem 5.6 we can obtain $Dg(\boldsymbol{\theta})$ using (5.10) and (5.11), which tells the effects of changes in t, G, M and P on output Y and interest rate r. If the purpose of the analysis is to study the impact of one exogenous variable on one endogenous variable only, then we can use Cramer's rule instead of inverting $D_{\mathbf{x}} f(\mathbf{x}, \boldsymbol{\theta})$. For example, to find the impact of a tax increase on output, we replace the first column of $D_{\mathbf{x}} f(\mathbf{x}, \boldsymbol{\theta})$ with the first column of $D_{\boldsymbol{\theta}} f(\mathbf{x}, \boldsymbol{\theta})$, divide the determinant with J and add a negative sign:

$$\begin{aligned}
\frac{\partial Y}{\partial t} &= -\frac{1}{J} \begin{vmatrix} C'[\cdot]Y & -\partial I/\partial r \\ 0 & -\partial L/\partial r \end{vmatrix} \\
&= \frac{C'[\cdot]Y(\partial L/\partial r)}{J} < 0.
\end{aligned}$$

5.4   Exercises

1. Let f be a linear operator on $\mathbb{R}^n$. Find the derivative of f by using the partial derivatives in Theorem 5.1.
2. Find the derivative $Df(\mathbf{x})$ of the function $f : \mathbb{R}^2 \to \mathbb{R}^2$, defined by

$$f(\mathbf{x}) = (x_1 + \sin x_2, x_2 e^{x_1}).$$

3. Find $\nabla f(\mathbf{x})$ and $\nabla^2 f(\mathbf{x})$ in Eq. (5.3).
4. Find the Hessian $\nabla^2 f(\mathbf{x})$ of the function $f : \mathbb{R}^2_{++} \to \mathbb{R}$ at $x = (2, 1)$ such that

$$f(\mathbf{x}) = x_1 + \sqrt{x_2}.$$

Determinate the definiteness of $\nabla^2 f(\mathbf{x})$ at that point.
5. Let $f : \mathbb{R}^2_{++} \to \mathbb{R}$, $f(\mathbf{x}) = (x_1 x_2)^{1/2}$.
   (a) Find the gradient $\nabla f(\mathbf{x})$ at the point $\mathbf{x} = (1, 1)$.
   (b) What is the directional derivative $D_{\mathbf{u}} f(\mathbf{x})$ at $(1, 1)$ in the direction of the vector $\mathbf{u} = (2/\sqrt{5}, 1/\sqrt{5})$?

6. Suppose that f is a differentiable functional on $\mathbb{R}^n$. Suppose $\mathbf{u} \in \mathbb{R}^n$ is a unit vector, with $\|\mathbf{u}\| = 1$. Show that the directional derivative of f at $\mathbf{x}$ in the direction of $\mathbf{u}$ is

$$D_{\mathbf{u}} f(\mathbf{x}) = \nabla f(\mathbf{x})^{\mathrm{T}} \mathbf{u}.$$

7. Prove that the gradient of a differentiable functional f points in the direction of greatest increase. Hint: Use the Cauchy-Schwarz Inequality.
8. Suppose that $f(x) = x^2 - 1$.
   (a) Does f^{-1} exist in the neighbourhood of $x = 0$?
   (b) Find the second-order Taylor formula for f about the point $x = 0$.
9. Calculate the directional derivative of the function

$$f(x, y) = x + 2 \log y$$

at the point $(1, 2)$ in the direction $(1, 1)$.
10. Let $f : \mathbb{R}^2 \to \mathbb{R}$ be given by

$$f(x, y) = \frac{1}{1 + x^2 + y^2}.$$

   (a) Calculate the directional derivative of f at the point $(1, 0)$ in the direction of the vector $\mathbf{v} = (4, 3)$.
   (b) In which direction the value of f increases at the fastest rate? Explain.
11. Find the second-order Taylor formula for the following functions about the given point $\mathbf{x}_0$ in $\mathbb{R}^2$.
   (a) $f(\mathbf{x}) = (x_1 + x_2)^2$, $\mathbf{x}_0 = (0, 0)$
   (b) $f(\mathbf{x}) = x_1 x_2 + (x_1 + x_2)^3$, $\mathbf{x}_0 = (0, 0)$
   (c) $f(\mathbf{x}) = e^{x_1 + x_2}$, $\mathbf{x}_0 = (0, 0)$
   (d) $f(\mathbf{x}) = \sin(x_1 x_2) \cos x_2$, $\mathbf{x}_0 = (1, 0)$
12. Find the first-order and second-order Taylor approximations of the following functions at the indicated points:
   (a) $f(x) = \sin x$, at $x^* = 0$
   (b) $f(x) = \sin x$, at $x^* = \pi/2$
   (c) $f(x) = e^x$, at $x^* = 1$
   (d) $f(x) = \log(1 + x)$, at $x^* = 0$
13. In a dynamic growth model the relation between consumption in period $t + 1$, c_{t+1}, consumption in period t, c_t, and the capital stock in period $t + 1$, k_{t+1}, is governed by the Euler equation

$$\beta \frac{U'(c_{t+1})}{U'(c_t)} \left[F'(k_{t+1}) + 1 - \delta \right] = 1,$$

where U is a utility function, F is a production function, and β and δ are given parameters. Let $\mathbf{x} = [c_{t+1} \ c_t \ k_{t+1}]^{\mathrm{T}}$. Show that the first-order Taylor approximation of the Euler equation about the steady-state values $\mathbf{x}^* = [c^*\ c^*\ k^*]^T$ is

$$\beta\left[F'(k^*)+1-\delta+\frac{U''(c^*)}{U'(c^*)}[F'(k^*)+1-\delta](c_{t+1}-c_t)+F''(k^*)(k_{t+1}-k^*)\right] \simeq 1.$$

14. Prove the following log-linear approximations:
    (a) $x \simeq x^*(\hat{x}+1)$
    (b) $xy \simeq x^*y^*(\hat{x}+\hat{y}+1)$
    (c) $y_t = c_t + g_t$ can be represented by

$$\hat{y}_t = \frac{c^*}{y^*}\hat{c}_t + \frac{g^*}{y^*}\hat{g}_t.$$

15. Prove the following multivariate version of log-linear approximation:[3]
    Let $f : \mathbb{R}^n \to \mathbb{R}$ be a C^1 function. Then for any $\mathbf{x}^* = [x_1^*\ x_2^* \cdots x_n^*]^T \in \mathbb{R}^n$,

$$f(\mathbf{x}) \simeq f(\mathbf{x}^*) + \nabla f(\mathbf{x}^*)^T D\hat{\mathbf{x}},$$

where D is an $n \times n$ diagonal matrix with diagonal elements $x_1^*, x_2^*, \ldots, x_n^*$ and

$$\hat{\mathbf{x}} = [\log x_1 - \log x_1^*\ \ \log x_2 - \log x_2^* \cdots \log x_n - \log x_n^*]^T.$$

16. Let $f : \mathbb{R}^2 \to \mathbb{R}^2$ be given by

$$f(\mathbf{x}) = (x_1^2 - x_2^2,\ 2x_1x_2).$$

    (a) What is the range of f?
    (b) Is the Jacobian of f zero at any point in $\mathbb{R}^2$?
    (c) Put $\mathbf{x} = (2, 1)$ and let $\mathbf{y} = f(\mathbf{x})$. Compute $Df(\mathbf{x})$ and $Df^{-1}(\mathbf{y})$.
17. Let $f : \mathbb{R}^2 \to \mathbb{R}^2$ be given by

$$f(\mathbf{x}) = (x_1 \cos x_2,\ x_1 \sin x_2).$$

    Find the set of points that the inverse of f does not exist.
18. Let $f : \mathbb{R}^2 \to \mathbb{R}^2$ be the function

$$f(\mathbf{x}) = (e^{x_1} \cos x_2,\ e^{x_1} \sin x_2).$$

    (a) Find the derivative $Df(\mathbf{x})$.
    (b) Evaluate the Jacobian of f at the point $\mathbf{a} = (0, \pi/3)$.
    (c) Let $\mathbf{b} = f(\mathbf{a})$. Find the derivative $Df^{-1}(\mathbf{b})$.

---

[3]The idea is due to Shivaji Koirala.

19. Define the Hénon map $f : \mathbb{R}^2 \to \mathbb{R}^2$ by

$$f(\mathbf{x}) = (a - bx_2 - x_1^2,\ x_1),$$

where a and b are constants and $b \neq 0$.

(a) Show that $Df(\mathbf{x})$ is invertible at any point of $\mathbb{R}^2$. Thus every point of $\mathbb{R}^2$ has a neighbourhood in which f is a bijection.

(b) Find an explicit formula for f^{-1}. Derive $Df^{-1}(\mathbf{x})$, and verify the inverse function theorem.

The Hénon map exhibits a number of interesting properties in the study of complex dynamics. See Devaney (2003) for details.

20. Show that the equations

$$xy^5 + yu^5 + zv^5 = 1$$
$$x^5y + y^5u + z^5v = 1$$

have a unique solution $u = g_1(x, y, z)$, $v = g_2(x, y, z)$ in the neighbourhood of the point $(x, y, z, u, v) = (0, 1, 1, 1, 0)$. Find the matrix for $Dg(0, 1, 1)$, where $g = (g_1, g_2)$.

21. Consider the following Keynesian model in macroeconomics:

$$Y = C[(1 - t)(Y + B/P)] + I(r, Y) + G,$$
$$M/P = L(Y, r),$$

where Y (output) and r (interest rate) are endogenous and P (price), G (government expenditure), t (tax rate), M (money supply) and B (government bonds) are exogenous. C, I and L are functions for consumption, investment and money demand respectively. Using Cramer's rule, find the following partial derivatives in studying the comparative statics of the model.

(a) $\partial Y/\partial G$,

(b) $\partial r/\partial M$,

(c) $\partial Y/\partial B$.

What assumptions do you have to make in order for the model to work?

22. Suppose a consumer wants to maximize the following function with respect to consumption c_t and c_{t+1}:

$$V_t = U(c_t) + \beta U(c_{t+1}) \tag{5.12}$$

subject to the intertemporal budget constraint

$$c_t + \frac{c_{t+1}}{1 + r} = y_t + \frac{y_{t+1}}{1 + r}, \tag{5.13}$$

where U is the instantaneous utility function, β is a discount factor, y_t and y_{t+1} are incomes in periods t and $t+1$ respectively, and r is the interest rate.

(a) By writing (5.12) as

$$f(x, \theta) = U(c_t) + \beta U(c_{t+1}) - V_t$$

with $x = c_{t+1}$ and $\theta = c_t$, apply the implicit function theorem to find the marginal rate of time preference, dc_{t+1}/dc_t.

(b) Find the slope dc_{t+1}/dc_t of the intertemporal budget constraint (5.13) as well.

(c) Combine your results to get the Euler equation:

$$\frac{\beta U'(c_{t+1})}{U'(c_t)}(1+r) = 1. \tag{5.14}$$

The Euler equation (5.14) is the cornerstone of the dynamic general equilibrium model used in macroeconomic analysis.

23. (Wickens 2011, chapter 8) In the transaction cost approach to money demand, households incur a transaction cost in consumption. The following two conditions are satisfied in the steady state:

$$c + \pi m + T(c, m) - x - \theta b = 0, \tag{5.15}$$

$$T_m(c, m) + R = 0, \tag{5.16}$$

where c is consumption, m is money demand, x is exogenous income, b is bond holding and R is the market interest rate. The transaction cost T is a function of c and m, which has the following properties:

$$T(0, m) = 0, \ T_c \geq 0, \ T_{cc} \geq 0, \ T_m \leq 0, \ T_{mm} > 0, \ T_{mc} = 0,$$

where T_c means $\partial T/\partial c$, and T_{mc} means $\partial^2 T/\partial m \partial c$, etc. Equations (5.15) and (5.16) can be written as $f(\mathbf{x}, \mathbf{y}) = \mathbf{0}$, where $f : \mathbb{R}^2 \times \mathbb{R}^3 \rightarrow \mathbb{R}^2$, $\mathbf{x} = (c, m)^T$ is the vector of endogenous variables, and $\mathbf{y} = (x, b, R)^T$ are exogenous.

(a) Find the impact of an increase in interest rate on money demand.

(b) Find the effect of a decrease in exogenous income x on consumption.

References

Brosowski, B., & Deutsch, F. (1981). An elementary proof of the Stone-Weierstrass theorem. *Proceedings of the American Mathematical Society, 81*(1), 89–92.

Clausen, A. (2006). *Log-linear approximations*. Unpublished class note, University of Pennsylvania.

Devaney, R. L. (2003). *An introduction to chaotic dynamical systems*, Second Edition. Cambridge: Westview Press.

Rudin, W. (1976). *Principles of mathematical analysis*, Third edition. New York: McGraw-Hill.

Wickens, M. (2011). *Macroeconomic theory: A dynamic general equilibrium approach*, Second Edition. Princeton: Princeton University Press.

Convex Analysis

6

Models in economic analysis often assume sets or functions are convex. These convexity properties make solutions of optimization problem analytically convenient. In this chapter we discuss various concepts of convexity in Euclidean spaces and their implications. The readers can consult Rockafellar (1970) for more details.

6.1   Basic Definitions

Suppose that $\mathbf{x}, \mathbf{y}$ are vectors in an n-dimensional vector space V, where $n \geq 2$. If $\mathbf{x}$ and $\mathbf{y}$ are linearly independent, then the set of all linear combinations of the two vectors,

$$S = \operatorname{span}(\{\mathbf{x}, \mathbf{y}\}) = \{\alpha\mathbf{x} + \beta\mathbf{y} : \alpha, \beta \in \mathbb{R}\}$$

is a two-dimensional subspace of V. If we impose the restriction $\beta = 1 - \alpha$, the set

$$M = \{\alpha\mathbf{x} + (1 - \alpha)\mathbf{y} : \alpha \in \mathbb{R}\}$$

is called an **affine set** and is no longer two dimensional or a subspace. To see this, we can express a vector $\mathbf{z} \in M$ as

$$\mathbf{z} = \alpha\mathbf{x} + (1 - \alpha)\mathbf{y} = \mathbf{y} + \alpha(\mathbf{x} - \mathbf{y}),$$

for some $\alpha \in \mathbb{R}$. Geometrically, $\mathbf{x} - \mathbf{y}$ is the vector going from the head of $\mathbf{y}$ towards the head of $\mathbf{x}$. Therefore $\mathbf{z}$ is a vector on the straight line passing through the heads of $\mathbf{x}$ and $\mathbf{y}$. Since this line in general does not include the zero vector, M is not a subspace of V. A vector in M is called an **affine combination** of $\mathbf{x}$ and $\mathbf{y}$.

Now we further restrict the value of α to be between 0 and 1. Then the set

$$C = \{\alpha\mathbf{x} + (1 - \alpha)\mathbf{y} : \alpha \in [0, 1]\}$$

© Springer Nature Switzerland AG 2019
K. Yu, *Mathematical Economics*, Springer Texts in Business and Economics,
https://doi.org/10.1007/978-3-030-27289-0_6

is called the **convex combinations** of **x** and **y**. It is clear that this is the set of vectors lying between the heads of **x** and **y**. The idea can be extended to more than two vectors. Let $A = \{\mathbf{x}_1, \mathbf{x}_2, \ldots, \mathbf{x}_n\}$ be a set of vectors in V, then the convex combinations of A is defined as the set

$$C = \left\{ \sum_{i=1}^{n} \alpha_i \mathbf{x}_i : \alpha_i \geq 0, i = 1, \ldots, n, \sum_{i=1}^{n} \alpha_i = 1 \right\}. \tag{6.1}$$

6.2   Convex Sets

In Chap. 3 we define a convex set in $\mathbb{R}^n$ under the discussion of a metric space. In general, a set S in a vector space V is convex if for any $\mathbf{x}, \mathbf{y} \in S$, the convex combinations of $\mathbf{x}$ and $\mathbf{y}$ are in S. Otherwise the set is called non-convex (see Fig. 6.1). The **convex hull** of a set $A \subseteq V$ is defined to be the set of all convex combinations of vectors in A and is denoted by conv(A). That is,

$$\text{conv}(A) = \{\alpha \mathbf{x} + (1 - \alpha)\mathbf{y} : \mathbf{x}, \mathbf{y} \in A, \alpha \in [0, 1]\}.$$

The convex hull of A is the smallest convex set in V that contains A. In other words, conv(A) is the intersection of all convex sets that contain A. When A is a finite set of say n vectors, conv(A) is the set defined in Eq. (6.1).

Example 6.1 Let $\mathcal{B} = (\mathbf{e}_1, \mathbf{e}_2, \ldots, \mathbf{e}_n)$ be the standard basis of $\mathbb{R}^n$. Then the convex hull of $\mathcal{B}$ is called a **standard simplex** in $\mathbb{R}^n$.

Example 6.2 In consumer theory, the preference relation of a consumer is often assumed to be convex. More precisely, the upper contour set, also called the no-worse-than set, of a consumption bundle **a**, $\succsim(\mathbf{a})$, is a convex set.

Theorem 6.1 *Let A and B be subsets of a vector space V. Then the convex hull of their Minkowski sum is the Minkowski sum of their convex hulls. That is,*

Fig. 6.1 Convex and non-convex sets

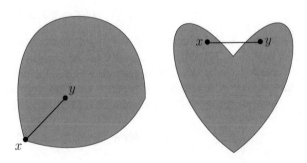

$$conv(A + B) = conv(A) + conv(B).$$

The proof is left as an exercise. By the principle of mathematical induction, the theorem applies to any finite number of set, that is, for any sets $A_1, A_2, \ldots, A_n$,

$$\text{conv}\left(\sum_{i=1}^{n} A_i\right) = \sum_{i=1}^{n} \text{conv}(A_i). \tag{6.2}$$

An important result in convex analysis is the **Shapley-Folkman theorem**, which roughly says that the sum of a large number of sets in an Euclidean space, not necessary convex, is approximately convex. That is, the sum of sets $\sum_{i=1}^{n} A_i$ for large n converges to the sum in Eq. (6.2). See Starr (2008) for a formal statement of the theorem.

Example 6.3 We define the aggregate production set of an economy in Example 4.9 as

$$Y = Y_1 + Y_2 + \cdots + Y_l = \sum_{j=1}^{l} Y_j.$$

The Shapley-Folkman theorem implies that even though the production sets Y_i of individual firms are not convex, the aggregate production set Y can be assumed to be convex, making the analysis more tractable.

Next we turn to the idea of a cone. A set S in a vector space V is called a **cone** if for any $\mathbf{x} \in S$ and $\alpha \geq 0$, the vector $\alpha\mathbf{x}$ is in S. By definition a cone must contain the zero vector. A convex cone, as the name implies, is a cone that is also convex. Formally, a set $S \in V$ is a **convex cone** if for any $\mathbf{x}, \mathbf{y} \in S$ and $\alpha, \beta \geq 0$, $\alpha\mathbf{x} + \beta\mathbf{y} \in S$.

Example 6.4 Suppose that $f : \mathbb{R} \to \mathbb{R}$ is given by

$$f(x) = |ax|,$$

for $a \in \mathbb{R}$. Then the epigraph of f is a convex cone in $\mathbb{R}^2$.

Example 6.5 In consumer theory, the consumption set of a consumer $\mathbb{R}_+^n$ is a convex cone. In production theory, if a firm's technology exhibits constant returns to scale, then the production set is a convex cone in $\mathbb{R}^n$.

6.3   Affine Sets

6.3.1   Definitions

An affine set M is a subset of a vector space V such that for every $\mathbf{x}, \mathbf{y} \in M$ and $\alpha \in \mathbb{R}$,

$$\alpha \mathbf{x} + (1 - \alpha)\mathbf{y} \in M.$$

The empty set $\varnothing$ and V are trivial examples of affine sets, and so is the set $\{\mathbf{x}\}$ which contains any single vector $\mathbf{x} \in V$. In general, affine sets are "linear" sets such as a straight line or a plane in an Euclidian space.

Theorem 6.2 *An affine set which contains the zero vector is a subspace.*

Proof It is clear that if $S \in V$ is a subspace, then it is an affine set with the zero vector. Conversely, let M be an affine set which contains $\mathbf{0}$. For M to be a subspace, we need to show that it is closed under addition and scalar multiplication. Let $\mathbf{x}, \mathbf{y} \in M$. Then for any $\alpha \in \mathbb{R}$,

$$\alpha \mathbf{x} = \alpha \mathbf{x} + (1 - \alpha)\mathbf{0} \in M.$$

Also,

$$\frac{1}{2}(\mathbf{x} + \mathbf{y}) = \frac{1}{2}\mathbf{x} + \left(1 - \frac{1}{2}\right)\mathbf{y} \in M$$

so that

$$\mathbf{x} + \mathbf{y} = 2\left(\frac{1}{2}(\mathbf{x} + \mathbf{y})\right) \in M$$

as required.   □

Given an affine set M, the **translation** of M by a vector $\mathbf{a}$ is defined as

$$M + \mathbf{a} = \{\mathbf{x} + \mathbf{a} : \mathbf{x} \in M\}.$$

It is easy to verify that $M + \mathbf{a}$ is also an affine set. In fact we say an affine set L is **parallel** to an affine set M if $L = M + \mathbf{a}$ for some $\mathbf{a} \in V$. Theorem 6.2 implies that every nonempty affine set M is parallel to a unique subspace S, given by

$$S = \{\mathbf{x} - \mathbf{a} : \mathbf{x}, \mathbf{a} \in M\}.$$

6.3.2 Hyperplanes and Half-Spaces

Let V be an n-dimensional vector space. Recall that the kernel of a linear functional f on V is a subspace of dimension $n-1$, which we called a hyperspace in Sect. 4.3.3. The hyperplane we defined in (4.8) is in fact a translation of this hyperspace. We can look at hyperplanes again in the context of an inner product space. Let S be a one-dimensional subspace of V. Every basis of S consists of a single nonzero vector, say, $\mathbf{b}$. The orthogonal complement of S is a subspace of dimension $n-1$:

$$S^{\perp} = \{\mathbf{x} \in V : \mathbf{b}^{\mathrm{T}}\mathbf{x} = 0\}. \tag{6.3}$$

A hyperplane is an affine set parallel to $S^{\perp}$, that is, for any vector $\mathbf{a} \in V$,

$$\begin{aligned} S^{\perp} + \mathbf{a} &= \{\mathbf{x} + \mathbf{a} : \mathbf{b}^{\mathrm{T}}\mathbf{x} = 0\} \\ &= \{\mathbf{y} : \mathbf{b}^{\mathrm{T}}(\mathbf{y} - \mathbf{a}) = 0\} \tag{6.4} \\ &= \{\mathbf{y} : \mathbf{b}^{\mathrm{T}}\mathbf{y} = c\} \tag{6.5} \end{aligned}$$

where $c = \mathbf{b}^{\mathrm{T}}\mathbf{a}$. For example, by taking $\mathbf{b} = (p_1, \ldots, p_n)$ in $\mathbb{R}^n$ and $c = \pi$, the hyperplane in (6.5) becomes the isoprofit hyperplane in (4.9). Notice that in (6.4) the hyperplane is orthogonal to the vector $\mathbf{b}$ and passes through the vector $\mathbf{a}$.

For any nonzero vector $\mathbf{b} \in V$ and any $c \in \mathbb{R}$, two closed **half-spaces** can be defined as

$$H^{+} = \{\mathbf{x} \in V : \mathbf{b}^{\mathrm{T}}\mathbf{x} \geq c\}$$

and

$$H^{-} = \{\mathbf{x} \in V : \mathbf{b}^{\mathrm{T}}\mathbf{x} \leq c\}.$$

All half-spaces are convex sets. If $c = 0$, then the half-space becomes a convex cone. Obviously, $H^{+} \cap H^{-}$ is a hyperplane. Half-spaces are useful in defining feasible sets in economics. For example, the budget set B of a consumer can be represented by the intersection of $n + 1$ half-spaces:

$$B = \{\mathbf{x} : \mathbf{p}^{\mathrm{T}}\mathbf{x} \leq M\} \cap \{\mathbf{x} : \mathbf{e}_1^{\mathrm{T}}\mathbf{x} \geq 0\} \cap \cdots \cap \{\mathbf{x} : \mathbf{e}_n^{\mathrm{T}}\mathbf{x} \geq 0\}$$

where $\mathbf{p}$ is the market price vector of the n goods and M is the income of the consumer. Notice that the intersections of any number of half-space is a convex set.

6.4   Convex Functions

6.4.1   Convex and Concave Functions

Let $f : S \to \mathbb{R}$ be a functional on a subset S of an inner product space V. Recall that the epigraph of the function is defined as

$$\text{epi } f = \{(\mathbf{x}, y) \in S \times \mathbb{R} : f(\mathbf{x}) \le y\}.$$

The hypograph of f, on the other hand, is defined as

$$\text{hypo } f = \{(\mathbf{x}, y) \in S \times \mathbb{R} : f(\mathbf{x}) \ge y\}.$$

It is obvious that the intersection of epi f and hypo f is the graph of f. We call f a **convex function** if epi f is a convex subset of $V \times \mathbb{R}$. The definition implies that the domain S of f is a convex set in V. If hypo f is a convex subset of $V \times \mathbb{R}$, then f is called a **concave function**. A function is called an **affine function** if it is both convex and concave. It is straightforward to show that f is convex if and only if $-f$ is concave. An important characterization of convex functions is as follows.

Theorem 6.3 *A function $f : S \to \mathbb{R}$ is convex if and only if for every $\mathbf{x}, \mathbf{z} \in S$ and $0 < \alpha < 1$,*

$$f((1 - \alpha)\mathbf{x} + \alpha\mathbf{z}) \le (1 - \alpha)f(\mathbf{x}) + \alpha f(\mathbf{z}). \tag{6.6}$$

Proof Suppose that f is a convex function on S. By definition $(\mathbf{x}, f(\mathbf{x}))$ and $(\mathbf{z}, f(\mathbf{z}))$ are in epi f. Since epi f is convex, for $0 < \alpha < 1$,

$$((1 - \alpha)\mathbf{x} + \alpha\mathbf{z}, (1 - \alpha)f(\mathbf{x}) + \alpha f(\mathbf{z})) = (1 - \alpha)(\mathbf{x}, f(\mathbf{x})) + \alpha(\mathbf{z}, f(\mathbf{z})) \in \text{epi } f,$$

which implies (6.6). The converse is straightforward and is left as an exercise.   □

Theorem 6.3 can be generalized to any convex combination:

Theorem 6.4 (Jensen's Inequality) *A function $f : S \to \mathbb{R}$ is convex if and only if for every $\mathbf{x}_1, \dots, \mathbf{x}_m \in S$ and $\alpha_1 \ge 0, \dots, \alpha_m \ge 0$ with $\sum_{i=1}^{m} \alpha_i = 1$,*

$$f(\alpha_1\mathbf{x}_1 + \cdots + \alpha_m\mathbf{x}_m) \le \alpha_1 f(\mathbf{x}_1) + \cdots + \alpha_m f(\mathbf{x}_m).$$

The inequalities in Theorems 6.3 and 6.4 are reversed if f is a concave function. The inequalities turn into equalities if f is an affine function. In fact, every affine function can be expressed in the form

$$f(\mathbf{x}) = \mathbf{a}^\mathsf{T}\mathbf{x} + b$$

where $\mathbf{a} \in V$ and $b \in \mathbb{R}$.

Example 6.6 The following are some examples of convex function of single variable, which can be characterized by $f''(x) \geq 0$.

1. $f(x) = e^{ax}, a \in \mathbb{R}$;
2. $f(x) = x^a$, for $x \geq 0$ and $a \geq 1$;
3. $f(x) = -\log x$, for $x > 0$.

Example 6.7 Applying Theorem 6.4 to the negative log function above, we have

$$-\log(\alpha_1 x_1 + \cdots + \alpha_m x_m) \leq -\alpha_1 \log(x_1) - \cdots - \alpha_m \log(x_m).$$

Multiplying both sides by -1 and taking the exponential, we have

$$\alpha_1 x_1 + \cdots + \alpha_m x_m \geq x_1^{\alpha_1} \cdots x_m^{\alpha_m}. \tag{6.7}$$

Setting $\alpha_i = 1/m$ for $i = 1, \ldots, m$, the inequality in (6.7) shows that the arithmetic mean of m positive numbers is bounded below by their geometric mean.

New convex functions can be constructed using simple convex functions. The following results are useful for this purpose and for recognizing convex functions.

1. Suppose f is a convex function on a convex set S in a vector space V and $g : \mathbb{R} \to \mathbb{R}$ is an increasing and convex function. Then $g \circ f$ is also a convex function. For example, if f is convex, then $h(\mathbf{x}) = e^{f(\mathbf{x})}$ is convex.
2. If f and g are convex functions on a convex set S, then $\alpha f + g$ is also convex for $\alpha > 0$. Consequently, the set of all convex functions on S is a cone.
3. Let T be a convex set in $\mathbb{R}^{n+1}$, and let S be the orthogonal projection of T on $\mathbb{R}^n$. For every $\mathbf{x} \in S$, define

$$f(\mathbf{x}) = \inf\{y : (\mathbf{x}, y) \in T\}.$$

Then f is a convex function on S. This result is like the converse of the definition of a convex function. Here T is contained in the epigraph of f. The orthogonal projection of T on $\mathbb{R}^n$ becomes the domain of f. Figure 6.2 illustrates an example for $n = 2$. The underside of T is the graph of the convex function f.

4. Suppose $f : V \to W$ is a linear transformation and g is a convex function on W. Then $g \circ f$ is a convex function on V.

The following theorem lists the different characterizations of a concave function.

Theorem 6.5 *Suppose f is a C^2 functional on an open convex set $S \in \mathbb{R}^n$. The following statements are equivalent:*

1. Function f is concave.
2. Function $-f$ is convex.

Fig. 6.2 Constructing a
convex function from a
convex set

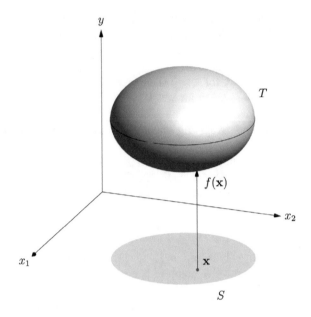

3. *The hypograph of* f, $\{(\mathbf{x}, z) \in S \times \mathbb{R} : f(\mathbf{x}) \geq z\}$ *is convex.*
4. *For every* $\mathbf{x}, \mathbf{y} \in S$ *and* $\alpha \in [0, 1]$,

$$f(\alpha\mathbf{x} + (1 - \alpha)\mathbf{y}) \geq \alpha f(\mathbf{x}) + (1 - \alpha)f(\mathbf{y}).$$

5. *For every* $\mathbf{x}, \mathbf{y} \in S$,

$$(\nabla f(\mathbf{x}) - \nabla f(\mathbf{y}))^{\mathrm{T}}(\mathbf{x} - \mathbf{y}) \leq 0.$$

6. *For every* $\mathbf{x}, \mathbf{x}_0 \in S$,

$$f(\mathbf{x}) \leq f(\mathbf{x}_0) + \nabla f(\mathbf{x}_0)^{\mathrm{T}}(\mathbf{x} - \mathbf{x}_0).$$

7. *For every* $\mathbf{x} \in S$, *the Hessian* $\nabla^2 f(\mathbf{x})$ *is negative semi-definite.*

6.4.2   Quasi-Convex and Quasi-Concave Functions

Economists sometimes find that convex and concave functions are too restrictive.
A more general class of functions includes the quasi-convex and quasi-concave
functions.

Let $f : S \to \mathbb{R}$ be a functional on a subset S of an inner product space V. Then
f is called a **quasi-convex function** if, for $0 \leq \alpha \leq 1$ and $\mathbf{x}, \mathbf{z} \in S$,

$$f((1 - \alpha)\mathbf{x} + \alpha\mathbf{z}) \leq \max\{f(\mathbf{x}), f(\mathbf{z})\}.$$

Similarly, f is called a **quasi-concave function** if, for $0 \leq \alpha \leq 1$ and $\mathbf{x}, \mathbf{z} \in S$,

$$f((1 - \alpha)\mathbf{x} + \alpha\mathbf{z}) \geq \min\{f(\mathbf{x}), f(\mathbf{z})\}.$$

Quasi-convex and quasi-concave functions are less restrictive than convex and concave functions. Some properties of these functions are listed below. More can be found in Carter (2001, p. 336–342).

1. If f is concave (convex), then it is quasi-concave (convex).
2. Monotone functions are both quasi-convex and quasi-concave.

The following theorem lists the characterizations of quasi-concave functions.

Theorem 6.6 *Suppose f is a C^2 functional on an open convex set $S \in \mathbb{R}^n$. The following statements are equivalent:*

1. *Function f is quasi-concave.*
2. *Function $-f$ is quasi-convex.*
3. *For all $\mathbf{x}, \mathbf{y} \in S$ and $0 < \alpha < 1$,*

$$f(\alpha\mathbf{x} + (1 - \alpha)\mathbf{y}) \geq \min\{f(\mathbf{x}), f(\mathbf{y})\}.$$

4. *For all $\mathbf{a} \in S$, the upper contour set $\succsim_f(\mathbf{a}) = \{\mathbf{x} \in S : f(\mathbf{x}) \geq f(\mathbf{a})\}$ is a convex set.*
5. *For all $\mathbf{x}, \mathbf{y} \in S$, $f(\mathbf{y}) \geq f(\mathbf{x})$ implies that $\nabla f(\mathbf{x})^{\mathrm{T}}(\mathbf{y} - \mathbf{x}) \geq 0$.*
6. *For all $\mathbf{x} \in S$ and $\mathbf{u}^{\mathrm{T}}\mathbf{u} = 1$, $\nabla f(\mathbf{x})^{\mathrm{T}}\mathbf{u} = 0$ implies that*
   (a) $\mathbf{u}^{\mathrm{T}}\nabla^2 f(\mathbf{x})\mathbf{u} < 0$, or
   (b) $\mathbf{u}^{\mathrm{T}}\nabla^2 f(\mathbf{x})\mathbf{u} = 0$ and $g(t) = f(\mathbf{x} + t\mathbf{u})$ is a quasi-concave function of one variable of t such that $\mathbf{x} + t\mathbf{u} \in S$.

Figure 6.3 illustrates the classification of curvatures for the simple case of single variable functions. Parts (a) and (b) are strictly concave functions but (b) is monotone (increasing in this case). Part (c) is an affine function and therefore both concave and convex. Parts (d) and (e) are strictly convex with (d) being monotone. All except (e), however, are quasi-concave functions. Similarly, all except (a) are quasi-convex functions. Part (f) is neither convex nor concave but since it is monotone it is both quasi-convex and quasi-concave.

6.4.3 Increasing Functions on $\mathbb{R}^n_+$

If we want to define an increasing function $f : \mathbb{R}^n_+ \to \mathbb{R}_+$, we first need an order in the domain. A partial order on $\mathbb{R}^n_+$ can be defined as follows. For any $\mathbf{x}, \mathbf{y} \in \mathbb{R}^n_+$,

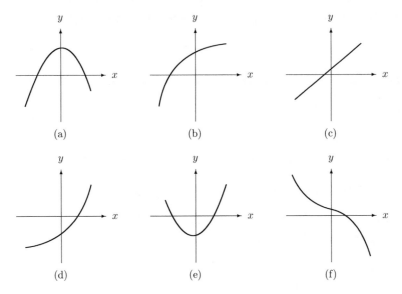

Fig. 6.3 Convex and concave functions

$\mathbf{x} \geq \mathbf{y}$ if $x_i \geq y_i$ for $i = 1, \ldots, n$. In economics, we frequently use the following induced relations on $\mathbb{R}^n_+$ as well:

1. $\mathbf{x} > \mathbf{y}$ if $\mathbf{x} \geq \mathbf{y}$ and $\mathbf{x} \neq \mathbf{y}$,
2. $\mathbf{x} \gg \mathbf{y}$ if $x_i > y_i$ for $i = 1, \ldots, n$.

For example, in consumer analysis, the non-satiation axiom implies that the utility function U of a consumer is increasing. That is, if $\mathbf{x}$ and $\mathbf{y}$ are two consumption bundles and $\mathbf{x} \geq \mathbf{y}$, then $U(\mathbf{x}) \geq U(\mathbf{y})$.

6.5 Homogeneous Functions

Let S be a cone in a vector space V. A functional f on S is said to be **homogeneous of degree k** if, for every $\alpha > 0$ and $\mathbf{x} \in S$,

$$f(\alpha \mathbf{x}) = \alpha^k f(\mathbf{x}).$$

If $k = 1$, then f is called a linearly homogeneous function. For example, a Cobb-Douglas function of two variables with a general form

$$f(x, y) = ax^\beta y^\gamma$$

is homogeneous of degree $\beta + \gamma$.

A functional f on a convex cone S is called **homothetic** if, for every $\alpha > 0$ and $\mathbf{x}, \mathbf{z} \in S$,

$$f(\mathbf{x}) = f(\mathbf{z})$$

implies that

$$f(\alpha \mathbf{x}) = f(\alpha \mathbf{z}).$$

The order $\succsim$ on S induced by a homothetic function is often called a homothetic preference relation. It follows that if $\mathbf{x} \sim \mathbf{z}$, then $\alpha \mathbf{x} \sim \alpha \mathbf{z}$.

In economics we frequently encounter linearly homogeneous functions. For example, in production theory, a cost function $C(\mathbf{p}, y)$, where $\mathbf{p}$ is the price vector of the input factors and y is the quantity of output, is linearly homogeneous in $\mathbf{p}$.[1] That is,

$$C(\alpha \mathbf{p}, y) = \alpha C(\mathbf{p}, y).$$

All linear functions are linearly homogeneous but the converse is not true. For example, the Cobb-Douglas function, $f(\mathbf{x}) = \prod_i x_i^{\beta_i}$, $\beta_i > 0$, $\sum_i \beta_i = 1$, is linearly homogeneous but not linear. Nevertheless, linearly homogeneous functions have properties that closely resemble linear functions. The following result illustrates this point.

Theorem 6.7 (Euler's Theorem) *Suppose that* $f : \mathbb{R}_+^n \to \mathbb{R}$ *is a linearly homogeneous* C^2 *functional, that is,*

$$f(\alpha \mathbf{x}) = \alpha f(\mathbf{x}) \tag{6.8}$$

for $\alpha > 0$. *Then for all* $\mathbf{x} \in \mathbb{R}_{++}^n$,

1. $f(\mathbf{x}) = \nabla f(\mathbf{x})^{\mathsf{T}} \mathbf{x}$,
2. $\nabla^2 f(\mathbf{x}) \mathbf{x} = \mathbf{0}$.

Proof Differentiating both sides of (6.8) with respect to α we have

$$[\nabla f(\alpha \mathbf{x})]^{\mathsf{T}} \mathbf{x} = f(\mathbf{x}).$$

Setting $\alpha = 1$ gives the first result. Now differentiating both sides of (6.8) with respect to $\mathbf{x}$ gives

---

[1] In the context of consumer theory y is replaced by a given utility level u and C is often called an expenditure function.

$$\alpha \nabla f(\alpha \mathbf{x}) = \alpha \nabla f(\mathbf{x})$$

or

$$\nabla f(\alpha \mathbf{x}) = \nabla f(\mathbf{x}).$$

Now differentiate the above equation with respect to α to get

$$\nabla^2 f(\alpha \mathbf{x})\mathbf{x} = \mathbf{0}.$$

Setting $\alpha = 1$ gives the second result. $\square$

Applying the first part of Euler's theorem to the cost function, we have

$$C(\mathbf{p}, y) = [\nabla_{\mathbf{p}} C(\mathbf{p}, y)]^{\mathrm{T}} \mathbf{p}. \tag{6.9}$$

Since by definition $C(\mathbf{p}, y) = h(\mathbf{p}, y)^{\mathrm{T}} \mathbf{p}$, where $h : \mathbb{R}_{++}^{n+1} \to \mathbb{R}_{+}^{n}$ is the Hicksian demand function for the inputs, and (6.9) holds for any $\mathbf{p} \in \mathbb{R}_{++}^{n}$, we have $\nabla_{\mathbf{p}} C(\mathbf{p}, y) = h(\mathbf{p}, y)$, which is often called Shephard's Lemma. Applying the second part of Euler's theorem we have

$$\nabla_{\mathbf{p}}^2 C(\mathbf{p}, y)\mathbf{p} = \mathbf{0},$$

or

$$D_{\mathbf{p}} h(\mathbf{p}, y)\mathbf{p} = \mathbf{0}.$$

This can be written as

$$\sum_{j=1}^{n} \frac{\partial h_i(\mathbf{p}, y)}{\partial p_j} p_j = 0 \tag{6.10}$$

for $i = 1, \ldots, n$. Since each Hicksian demand function is decreasing in its own price, i.e. $\partial h_i(\mathbf{p}, y)/\partial p_i < 0$, Eq. (6.10) implies that at least one of $\partial h_i(\mathbf{p}, y)/\partial p_j$ must be positive. In other words, at least one of the other $n - 1$ inputs is a substitute for input i. In the case of $n = 2$, the two inputs normally are substitutes of each other.[2]

---

[2] An exception is the Leontief production function, $f(\mathbf{x}) = \min\{x_1/a_1, \ldots, x_n/a_n\}$, where $a_1, \ldots, a_n > 0$. In this case $\nabla_{\mathbf{p}}^2 C(\mathbf{p}, y)$ is a zero matrix. Note that f is not differentiable in this case.

6.6 Separating Hyperplanes

Suppose A and B are convex subsets of V. $H_f(c)$ is said to be a **separating hyperplane** of A and B if A is contained in one of the half-space defined by $H_f(c)$ and B is contained in the other half-space. In other words, for every $\mathbf{a} \in A$ and $\mathbf{b} \in B$,

$$f(\mathbf{a}) \leq c \leq f(\mathbf{b}) \quad \text{or} \quad f(\mathbf{b}) \leq c \leq f(\mathbf{a}).$$

If S is a convex subset of V, then $H_f(c)$ is called a **supporting hyperplane** of S if S is contained in one of the half-spaces of $H_f(c)$ and at least one vector of S lies in the hyperplane itself. That is, for every $\mathbf{x} \in S$,

$$f(\mathbf{x}) \geq c \quad \text{or} \quad f(\mathbf{x}) \leq c$$

and there exists $\mathbf{s} \in H_f(c) \cap S$. For example, in production theory, the cost minimization problem involves finding the supporting hyperplane of the upper contour set of the production function that is orthogonal to the input price vector $\mathbf{p}$. The following result is a form of separating hyperplane theorem.

Theorem 6.8 *Suppose S is a closed convex subset of a normed vector space V and $\mathbf{z} \notin S$. There exists a vector $\mathbf{p} \in V$ and a vector $\mathbf{s} \in S$ such that for every $\mathbf{x} \in S$,*

$$\mathbf{p}^T\mathbf{x} \geq \mathbf{p}^T\mathbf{s} > \mathbf{p}^T\mathbf{z}. \tag{6.11}$$

In Theorem 6.8, $\mathbf{s}$ is the point in S that is "closest" to $\mathbf{z}$ and $\mathbf{p}$ is the vector $\mathbf{s} - \mathbf{z}$. Then $\mathbf{z}$ and S are separated by a hyperplane orthogonal to $\mathbf{p}$ and passing through any point between $\mathbf{z}$ and $\mathbf{s}$. The hyperplane passing through $\mathbf{s}$, using the form expressed in (6.4), is

$$H_f(c) = \{\mathbf{y} : \mathbf{p}^T(\mathbf{y} - \mathbf{s}) = 0\}$$

where $c = \mathbf{p}^T\mathbf{s}$. In fact, (6.11) means that $H_f(c)$ is a supporting hyperplane of S, with the point $\mathbf{z}$ contained in the other half-space. It is important to emphasize that S must be convex, otherwise one of the half-space defined by $H_f(c)$ may not contain S.

Other relating results are listed below, in the context of a normed vector space V:

1. Minkowski's Theorem: A closed convex set is the intersection of its supporting half-spaces.
2. Two disjoint closed convex sets have a separating hyperplane.
3. Two closed convex sets which intersect at a single point only have a common separating hyperplane which is also a supporting hyperplane of both sets.

6.7   Exercises

1. Show that a set A in a vector space is convex if and only if $A = \text{conv}(A)$.
2. Let S and T be convex subsets of a vector space V.
   (a) Prove that $S + T$ is convex.
   (b) Prove that $S \cap T$ is convex.
   (c) Is $S \cup T$ convex?
3. Show that an open ball in $\mathbb{R}^n$ is convex.
4. Let $f : V \to W$ be a linear transformation. Suppose that $S \subseteq V$ and $T \subseteq W$ are convex sets. Show that $f(S)$ and $f^{-1}(T)$ are convex.
5. Show that the orthogonal projection of a convex set on a subspace is also convex.
6. Let $F(S, \mathbb{R})$ be the set of all functionals on a metric space S. Let $\mathcal{C}(S, \mathbb{R})$ be the subset of continuous functions in $F(S, \mathbb{R})$. Prove or disprove: $\mathcal{C}(S, \mathbb{R})$ is a convex set in $F(S, \mathbb{R})$.
7. Let $L(V)$ be the set of all linear operators on a finite-dimensional vector space V. Prove or disprove: The set of all invertible operators in $L(V)$ is a convex set.
8. Let A be a subset of a vector space V. Suppose that $0 \le \alpha \le 1$. Prove that A is convex if and only if $A = \alpha A + (1 - \alpha)A$.
9. Prove Theorem 6.1 on the convex hull of Minkowski sum.
10. When is a cone S in a vector space V a subspace of V? Is S a convex cone?
11. Prove the statement in Example 6.4.
12. Suppose S and T are convex cones in a vector space V. Show that $S + T$ is also a convex cone.
13. Let $F(X)$ be the set of all functionals on a set X. Show that the set of all increasing functionals on X is a cone in $F(X)$.
14. Let S be a subset of a vector space V.
    (a) Prove or disprove: If S is a convex cone, then it is a subspace of V.
    (b) Is the converse true? Explain.
15. The **conic hull** of a set A in a vector space V is the smallest cone in V that contains A, and is denoted by $\text{cone}(A)$.
    (a) Give a formal definition of the conic hull.
    (b) Let $V = \mathbb{R}^2$, $\mathbf{x} = (-1, 0)$, $\mathbf{y} = (-1, -1)$, and let $A = \text{conv}(\{\mathbf{x}, \mathbf{y}\})$. What is $\text{cone}(A)$?
    (c) Sketch the region of $\text{cone}(A) \in \mathbb{R}^2$ in a diagram.
16. Show that $S^\perp$ defined in (6.3) is a subspace of V.
17. Let $\mathbf{a}$ and $\mathbf{b}$ be two distinct vectors in a vector space V. Show that

$$A = \{\mathbf{x} : \mathbf{a}^\mathsf{T}\mathbf{x} \le 0\} \cap \{\mathbf{x} : \mathbf{b}^\mathsf{T}\mathbf{x} \le 0\}$$

    is a convex cone.
18. List the counter parts of Theorem 6.5 for a convex function.
19. Let $f : X \to \mathbb{R}$ be a convex function on a metric space X and $g : \mathbb{R} \to \mathbb{R}$ be a convex and increasing function. Show that $g \circ f$ is convex.

20. Let $f : \mathbb{R}^n_+ \to \mathbb{R}_+$ be an increasing and concave function. Show that if there exists a $\mathbf{z} \gg \mathbf{0}$ such that $f(\mathbf{z}) = 0$, then $f(\mathbf{x}) = 0$ for all $\mathbf{x} \in \mathbb{R}^n_+$. (Hint: First show that for all $\alpha \geq 0$, $f(\alpha \mathbf{z}) = 0$. Then for any $\mathbf{x} \in \mathbb{R}^n_+$, we can choose a large enough α such that $\alpha \mathbf{z} \gg \mathbf{x}$.)

21. Let f be a linearly homogeneous function on a convex cone S. Show that if f is convex, then for all $\mathbf{x}, \mathbf{z} \in S$,

$$f(\mathbf{x} + \mathbf{z}) \leq f(\mathbf{x}) + f(\mathbf{z}).$$

22. Let $f : \mathbb{R}^n_+ \to \mathbb{R}$ be a C^1 homogeneous function of degree k. Show that

$$kf(\mathbf{x}) = \nabla f(\mathbf{x})^{\mathrm{T}} \mathbf{x}.$$

23. Let S be a convex cone in $\mathbb{R}^n_+$. Suppose that $h : S \to \mathbb{R}$ is a linearly homogeneous function and $g : \mathbb{R} \to \mathbb{R}$ is a strictly increasing function. Show that $f = g \circ h$ is homothetic.

24. Prove Minkowski's Theorem.

25. Find a separating hyperplane of

$$S = \{(x_1, x_2) \in \mathbb{R}^2 : x_1 x_2 \geq 1, x_1 \leq 0, x_2 \leq 0\}$$

and the origin. Find the supporting hyperplane of S that is orthogonal to $(1, 1)$.

26. Suppose that the function $f : \mathbb{R}^2 \to \mathbb{R}$ is given by

$$f(x, y) = y - x^2.$$

   (a) Let $S = \succsim_f (0) = \{(x, y) : f(x, y) \geq 0\}$ be the upper contour set of the value 0 induced by f (see Sect. 2.5). Is S a convex set? Explain.
   (b) Find two supporting hyperplanes of S which pass through the point $\mathbf{z} = (-3, 8)$.

27. Let $L(V)$ be the set of linear operators on an n-dimensional vector space V. Show that the determinant function $D : L(V) \to \mathbb{R}$ is homogenous of degree n.

28. Suppose that f is a linear transformation from a vector space V to a vector space W. Show that the graph of f is a subspace in the product set $V \times W$.

References

Carter, M. (2001). *Foundations of mathematical economics*. Cambridge: The MIT Press.

Rockafellar, R. T. (1970). *Convex analysis*. Princeton: Princeton University Press.

Starr, R. M. (2008). Shapley-Folkman theorem. *The new Palgrave dictionary of economics*, Second Edition. Houndmills: Palgrave Macmillan.

Optimization

7

7.1 Unconstrained Optimization

In economics we frequently have to find the points that maximize or minimize a differentiable functional $f(\mathbf{x}, \boldsymbol{\theta})$. Here f is called the **objective function**, $\mathbf{x} \in \mathbb{R}^n$ is the **control** or endogenous variable, and $\boldsymbol{\theta} \in \mathbb{R}^l$ is the **parameter** or exogenous variable. The choice of $\mathbf{x}$ may be constrained to a subset $G(\mathbf{x}, \boldsymbol{\theta})$ of $\mathbb{R}^n$, which is often called the **feasible set**. In a maximization process, we write the problem as

$$\max_{\mathbf{x}} \quad f(\mathbf{x}, \boldsymbol{\theta})$$

$$\text{subject to} \quad \mathbf{x} \in G(\mathbf{x}, \boldsymbol{\theta}).$$

Let us ignore the parameter $\boldsymbol{\theta}$ for now. Suppose that $\mathbf{x}^*$ is a **local maximum** of f. That is, there exists a neighbourhood B containing $\mathbf{x}^*$ such that for every $\mathbf{x} \in G \cap B$,

$$f(\mathbf{x}^*) \geq f(\mathbf{x}). \tag{7.1}$$

The linear approximation of $f(\mathbf{x})$ about $\mathbf{x}^*$ is

$$f(\mathbf{x}) = f(\mathbf{x}^*) + \nabla f(\mathbf{x}^*)^{\mathrm{T}}(\mathbf{x} - \mathbf{x}^*). \tag{7.2}$$

Putting (7.2) into (7.1) gives

$$f(\mathbf{x}^*) \geq f(\mathbf{x}^*) + \nabla f(\mathbf{x}^*)^{\mathrm{T}}(\mathbf{x} - \mathbf{x}^*),$$

or

$$\nabla f(\mathbf{x}^*)^{\mathrm{T}}(\mathbf{x} - \mathbf{x}^*) \leq 0.$$

© Springer Nature Switzerland AG 2019
K. Yu, *Mathematical Economics*, Springer Texts in Business and Economics,
https://doi.org/10.1007/978-3-030-27289-0_7

Notice that the expression on the left-hand side of the above inequality is the directional derivative of f at $\mathbf{x}^*$ in the direction of $(\mathbf{x} - \mathbf{x}^*)$ multiplied by its length. If $\mathbf{x}^*$ is an interior point of G, then the inequality becomes an equality. Otherwise if there exists an $\mathbf{x}$ such that

$$\nabla f(\mathbf{x}^*)^{\mathrm{T}}(\mathbf{x} - \mathbf{x}^*) < 0,$$

then the directional derivative in the opposite direction is positive, contradicting the fact that $\mathbf{x}^*$ is a local maximum. Therefore the necessary condition for an unconstrained maximization is

$$\nabla f(\mathbf{x}^*) = \mathbf{0}. \tag{7.3}$$

Any point that satisfies (7.3) is called a **stationary point** of f. In the special case that the control variable is constrained to $\mathbb{R}^n_+$, the necessary condition can be summarized by the following complementary slackness condition:

$$\nabla f(\mathbf{x}^*) \leq \mathbf{0}, \quad \mathbf{x}^* \geq \mathbf{0}, \quad \nabla f(\mathbf{x}^*)^{\mathrm{T}}\mathbf{x}^* = 0.$$

Example 7.1 Suppose $f(x, y) = x^2 - y^2$. Then $\nabla f(x, y) = (2x \ -2y)^{\mathrm{T}}$ so that $\nabla f(0) = \mathbf{0}$. It is, however, obvious in Fig. 7.1 that the origin is neither a maximum nor a minimum (it is called a saddle point).

Fig. 7.1 Graph of the saddle $z = x^2 - y^2$

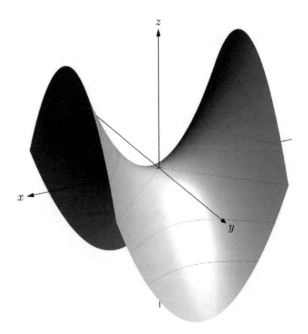

The above example shows that we need to find a sufficient condition for a local maximum. It turns out that such a condition exists for an interior **strict** local maximum. That is, $f(\mathbf{x}^*) > f(\mathbf{x})$ for all $\mathbf{x}$ in the neighbourhood of $\mathbf{x}^*$. The quadratic (Taylor) approximation of $f(\mathbf{x})$ about $\mathbf{x}^*$ is

$$f(\mathbf{x}) = f(\mathbf{x}^*) + \nabla f(\mathbf{x}^*)^{\mathrm{T}}(\mathbf{x} - \mathbf{x}^*) + \frac{1}{2}(\mathbf{x} - \mathbf{x}^*)^{\mathrm{T}}\nabla^2 f(\mathbf{x}^*)(\mathbf{x} - \mathbf{x}^*)$$

$$= f(\mathbf{x}^*) + \frac{1}{2}(\mathbf{x} - \mathbf{x}^*)^{\mathrm{T}}\nabla^2 f(\mathbf{x}^*)(\mathbf{x} - \mathbf{x}^*)$$

where in the first equality above $\nabla f(\mathbf{x}^*)^{\mathrm{T}}(\mathbf{x} - \mathbf{x}^*) = 0$ from the necessary condition in (7.3). Therefore $f(\mathbf{x}^*) > f(\mathbf{x})$ if and only if

$$(\mathbf{x} - \mathbf{x}^*)^{\mathrm{T}}\nabla^2 f(\mathbf{x}^*)(\mathbf{x} - \mathbf{x}^*) < 0.$$

This means that the Hessian $\nabla^2 f(\mathbf{x}^*)$ must be negative definite.

The above analysis deals with local maximum. If the objective function f is a concave function, then (7.3) becomes a necessary and sufficient condition for a **global maximum**.

7.2 Equality-Constrained Optimization

We want to solve the following problem:

$$\max_{\mathbf{x}} \quad f(\mathbf{x})$$

$$\text{subject to} \quad g(\mathbf{x}) = \mathbf{0}.$$

Here $f : \mathbb{R}^n \to \mathbb{R}$ is the objective function and $g : \mathbb{R}^n \to \mathbb{R}^m$ is the constraint function. Note that $m < n$ otherwise it is over constrained and $Dg(\mathbf{x})$ cannot have full rank m. The following result characterizes the necessary condition for a maximum.

Theorem 7.1 (Lagrange Multiplier Theorem) *Let $f : S \subseteq \mathbb{R}^n \to \mathbb{R}$ where S is open and $g : S \to \mathbb{R}^m$ be C^2 functions. Suppose $\mathbf{x}^*$ is a local maximum of f on the set $G = S \cap \{\mathbf{x} : g(\mathbf{x}) = \mathbf{0}\}$. Suppose also that $Dg(\mathbf{x}^*)$ has rank m. Then there exists a unique vector $\lambda \in \mathbb{R}^m$ such that*

$$\nabla f(\mathbf{x}^*) = Dg(\mathbf{x}^*)^{\mathrm{T}}\lambda.$$

In practice, we define a **Lagrangian function** $L : S \times \mathbb{R}^m \to \mathbb{R}$ such that

$$L(\mathbf{x}, \lambda) = f(\mathbf{x}) - g(\mathbf{x})^{\mathrm{T}}\lambda.$$

Then we set the gradients of L with respect to $\mathbf{x}$ and λ to zero as if we are maximizing L, that is,

$$\nabla_{\mathbf{x}} L(\mathbf{x}^*, \lambda) = \nabla f(\mathbf{x}^*) - Dg(\mathbf{x}^*)^{\mathrm{T}} \lambda = \mathbf{0} \qquad (7.4)$$

and

$$\nabla_\lambda L(\mathbf{x}^*, \lambda) = -g(\mathbf{x}) = \mathbf{0},$$

which give us the first-order conditions as in the Lagrange Multiplier Theorem.

The second-order conditions are a bit more complicated than those for the unconstrained optimization. First recall that the kernel of the linear function $Dg(\mathbf{x}^*)$ is defined as

$$K = \{\mathbf{h} \in \mathbb{R}^n : Dg(\mathbf{x}^*)\mathbf{h} = \mathbf{0}\}.$$

Geometrically, the kernel K is the linear approximation of the constraint $g(\mathbf{x}) = \mathbf{0}$ at the point $\mathbf{x}^*$. The following theorem states that the Hessian of L has something to do with the local curvature of f at $\mathbf{x}^*$ when $\mathbf{x}$ is constrained on this linear approximation of g. Since g is a vector-valued function, we have to express it in terms of its components in order to write down the second derivative. Hence we write g as $g(\mathbf{x}) = (g_1(\mathbf{x})\ g_2(\mathbf{x}) \cdots g_m(\mathbf{x}))^{\mathrm{T}}$. Then (7.4) can be written as

$$\nabla_{\mathbf{x}} L(\mathbf{x}^*, \lambda) = \nabla f(\mathbf{x}^*) - \sum_{i=1}^{m} \lambda_i \nabla g_i(\mathbf{x})$$

so that we can let $\nabla^2 L$ denote the $n \times n$ symmetric matrix

$$\nabla_{\mathbf{x}}^2 L(\mathbf{x}^*, \lambda) = \nabla^2 f(\mathbf{x}^*) - \sum_{i=1}^{m} \lambda_i \nabla^2 g_i(\mathbf{x}).$$

Theorem 7.2 *Suppose there exist points $\mathbf{x}^* \in G$ and $\lambda \in \mathbb{R}^m$ such that $Dg(\mathbf{x}^*)$ has rank m and $\nabla f(\mathbf{x}^*) = Dg(\mathbf{x}^*)^{\mathrm{T}} \lambda$.*

1. *If f has a local maximum on G at $\mathbf{x}^*$, then $\mathbf{h}^{\mathrm{T}} \nabla^2 L \mathbf{h} \leq 0$ for all $\mathbf{h} \in K$.*
2. *If $\mathbf{h}^{\mathrm{T}} \nabla^2 L \mathbf{h} < 0$ for all $\mathbf{h} \in K$ with $\mathbf{h} \neq \mathbf{0}$, then $\mathbf{x}^*$ is a strict local maximum of f on G.*

Note that in Theorem 7.2 part (a) is a necessary condition and part (b) is a sufficient condition. Also the definiteness of $\nabla^2 L$ can only be confirmed on the kernel K of $Dg(\mathbf{x}^*)$. The question now is to find a way to characterize the definiteness of $\nabla^2 L$ on K. For this we need the **bordered Hessian** at $\mathbf{x}^*$, which is defined as B_j for $j = 1, \ldots, n$:

$$B_j = \begin{pmatrix} 0 & \cdots & 0 & \partial g_1(\mathbf{x}^*)/\partial x_1 & \cdots & \partial g_1(\mathbf{x}^*)/\partial x_j \\ \vdots & \ddots & \vdots & \vdots & \ddots & \vdots \\ 0 & \cdots & 0 & \partial g_m(\mathbf{x}^*)/\partial x_1 & \cdots & \partial g_m(\mathbf{x}^*)/\partial x_j \\ \partial g_1(\mathbf{x}^*)/\partial x_1 & \cdots & \partial g_m(\mathbf{x}^*)/\partial x_1 & \partial^2 L(\mathbf{x}^*, \lambda)/\partial x_1^2 & \cdots & \partial^2 L(\mathbf{x}^*, \lambda)/\partial x_1 \partial x_j \\ \vdots & \ddots & \vdots & \vdots & \ddots & \vdots \\ \partial g_1(\mathbf{x}^*)/\partial x_j & \cdots & \partial g_m(\mathbf{x}^*)/\partial x_j & \partial^2 L(\mathbf{x}^*, \lambda)/\partial x_j \partial x_1 & \cdots & \partial^2 L(\mathbf{x}^*, \lambda)/\partial x_j^2 \end{pmatrix}.$$

For notational convenience, we let $Dg(\mathbf{x}^*)_j$ be the $m \times j$ matrix which includes the first j columns of $Dg(\mathbf{x}^*)$, and $\nabla^2 L_j$ be the $j \times j$ matrix which includes the first j rows and columns of $\nabla^2 L$, then

$$B_j = \begin{pmatrix} 0_{m \times m} & Dg(\mathbf{x}^*)_j \\ Dg(\mathbf{x}^*)_j^T & \nabla^2 L_j \end{pmatrix}.$$

Therefore B_j is a $(m + j) \times (m + j)$ symmetric square matrix, and B_n is the full bordered Hessian with the whole $Dg(\mathbf{x}^*)$ and $\nabla^2 L$.

Next, for the set of integers $N = \{1, 2, \ldots, n\}$, let π be a permutation of N. Then there are exactly $n!$ of such permutations. If we label the rows and columns of $\nabla^2 L$ by the numbers in N we can obtain $n!$ different matrices by applying a permutation π to the rows and columns at the same time. Similarly, π can be applied to the columns of $Dg(\mathbf{x}^*)$. Hence we denote $\nabla^2 L^\pi$ and $Dg(\mathbf{x}^*)^\pi$ the permutation matrices for the permutation π, and the permuted bordered Hessian

$$B_j^\pi = \begin{pmatrix} 0_{m \times m} & Dg(\mathbf{x}^*)_j^\pi \\ (Dg(\mathbf{x}^*)_j^\pi)^T & \nabla^2 L_j^\pi \end{pmatrix}$$

for $j = 1, \ldots, n$. Now we can state the second-order conditions for equality constrained maximization:

Theorem 7.3 *Let B_j be the bordered Hessian defined above. Then*

1. *$\mathbf{h}^T \nabla^2 L \mathbf{h} \leq 0$ for all $\mathbf{h} \in K$ if and only if for all permutation π of N and for $j = 1, \ldots, n$, we have $(-1)^j |B_j^\pi| \geq 0$,*
2. *$\mathbf{h}^T \nabla^2 L \mathbf{h} < 0$ for all $\mathbf{h} \in K$ with $\mathbf{h} \neq \mathbf{0}$ if and only if for $j = 1, \ldots, n$, we have $(-1)^j |B_j| > 0$.*

The proof of this theorem can be found in Debreu (1952).

Example 7.2 Find the maximum of the function $f(x, y) = x^2 - y^2$ in the constraint set $\{(x, y) \in \mathbb{R}^2 : x^2 + y^2 = 1\}$.

Here $n = 2$, $m = 1$, with $g(x, y) = x^2 + y^2 - 1$. The Lagrangian is $L(x, y, \lambda) = x^2 - y^2 - \lambda(x^2 + y^2 - 1)$. The first-order condition is

$$\nabla L(x, y, \lambda) = \begin{pmatrix} 2x - 2\lambda x \\ -2y - 2\lambda y \\ 1 - x^2 - y^2 \end{pmatrix} = \mathbf{0}.$$

The solutions to the equation are $(x^*, y^*, \lambda) = (1, 0, 1), (-1, 0, 1), (0, 1, -1)$ and $(0, -1, -1)$. Evaluating f at these four points gives $f(1, 0) = f(-1, 0) = 1$ and $f(0, 1) = f(0, -1) = -1$. Therefore $(1, 0)$ and $(-1, 0)$ are the maximum points. We can verify this by the second-order condition. Now

$$\nabla_{\mathbf{x}}^2 L(\mathbf{x}^*, \lambda) = \begin{pmatrix} 2 - 2\lambda & 0 \\ 0 & -2 - 2\lambda \end{pmatrix}$$

and $Dg(x^*, y^*) = (2x^*, 2y^*)$. The bordered Hessian matrices at the above four points are, respectively,

$$\begin{pmatrix} 0 & 2 & 0 \\ 2 & 0 & 0 \\ 0 & 0 & -4 \end{pmatrix}, \begin{pmatrix} 0 & -2 & 0 \\ -2 & 0 & 0 \\ 0 & 0 & -4 \end{pmatrix}, \begin{pmatrix} 0 & 0 & 2 \\ 0 & 4 & 0 \\ 2 & 0 & 0 \end{pmatrix}, \begin{pmatrix} 0 & 0 & -2 \\ 0 & 4 & 0 \\ -2 & 0 & 0 \end{pmatrix}.$$

Table 7.1 lists the values of $(-1)^j |B_j|$, $j = 1, 2$ for the four points, which confirms our results.

The function in the above example is called a saddle (see Fig. 7.1). The example in Sect. 7.1 shows that the function has no maximum nor minimum. Maxima and minima do exist, however, on the constrained set of the unit circle as demonstrated.

Example 7.3 In economics we frequently model a consumer's utility function with an increasing and quasi-concave C^2 function $u = f(\mathbf{x})$. If we assume an interior solution, the utility maximization problem becomes an equality constraint problem. The constraint of course is the budget hyperplane $g(\mathbf{x}) = \mathbf{p}^T\mathbf{x} - y$, where $\mathbf{p} \in \mathbb{R}_{++}^n$ is the price vector and $y > 0$ income. It follows from Table 5.1 that $\nabla g(\mathbf{x}) = \mathbf{p}$. Recall from Fig. 5.1 that the gradient of a functional is orthogonal to the tangent hyperplane of the level surface at any point. Therefore $\mathbf{p}$ is orthogonal to the budget hyperplane itself. Also, $\nabla f(\mathbf{x})$ is the vector orthogonal to tangent hyperplane of the level surface of f (in economics we call it an indifference surface) at any consumption bundle $\mathbf{x}$. The Lagrange Multiplier Theorem implies that at the optimal bundle $\mathbf{x}^*$,

$$\nabla f(\mathbf{x}^*) = \lambda \mathbf{p},$$

Table 7.1 Values of $(-1)^j |B_j|$

(x^*, y^*)	$-|B_1|$	$|B_2|$
$(1, 0)$	4	16
$(-1, 0)$	4	16
$(0, 1)$	0	-16
$(0, -1)$	0	-16

that is, $\nabla f(\mathbf{x}^*)$ and $\mathbf{p}$ are pointing in the same direction. Another way to say this is at $\mathbf{x}^*$, the budget hyperplane is tangent to the indifference surface, or, in the case of two goods, the marginal rate of substitution (MRS) is equal to the price ratio p_1/p_2.

□

Next we put the parameters $\boldsymbol{\theta}$ back into the objective function and the constraints. The maximum points are then a function of the parameters, $\mathbf{x}^* = \phi(\boldsymbol{\theta})$, so that $f(\mathbf{x}^*) = f(\phi(\boldsymbol{\theta}))$ which we call the **value function**. We want to study the effect on the value function when we change $\boldsymbol{\theta}$. In economics this is called comparative statics. The following result is very useful.

Theorem 7.4 (Envelope Theorem) *Given the maximization problem*

$$\max_{\mathbf{x}} \quad f(\mathbf{x}, \boldsymbol{\theta})$$

$$\textit{subject to} \quad g(\mathbf{x}, \boldsymbol{\theta}) = \mathbf{0}$$

where $\boldsymbol{\theta} \in \mathbb{R}^l$ is a vector of parameters. Suppose $\mathbf{x}^* \in G = \{\mathbf{x} : g(\mathbf{x}, \boldsymbol{\theta}) = 0\}$ is a local maximum. Then

$$\frac{\partial f(\mathbf{x}^*, \boldsymbol{\theta})}{\partial \theta_i} = \frac{\partial L(\mathbf{x}^*, \boldsymbol{\theta})}{\partial \theta_i}, \quad i = 1, \ldots, l$$

or in matrix form,

$$\nabla_{\boldsymbol{\theta}} f(\mathbf{x}^*, \boldsymbol{\theta}) = \nabla_{\boldsymbol{\theta}} L(\mathbf{x}^*, \boldsymbol{\theta})$$

where $L(\mathbf{x}^*, \boldsymbol{\theta}) = f(\mathbf{x}^*, \boldsymbol{\theta}) - g(\mathbf{x}^*, \boldsymbol{\theta})^{\mathrm{T}}\boldsymbol{\lambda}$ is the Lagrangian, and $\boldsymbol{\lambda} \in \mathbb{R}^m$ is the vector of Lagrange multipliers.

Proof Since $\mathbf{x}^*$ depends on the parameter $\boldsymbol{\theta}$ we can write $\mathbf{x}^* = \phi(\boldsymbol{\theta})$ where $\phi : \mathbb{R}^l \to \mathbb{R}^n$. Since $g(\mathbf{x}^*, \boldsymbol{\theta}) = 0$, $\boldsymbol{\lambda}^{\mathrm{T}} g(\mathbf{x}^*, \boldsymbol{\theta}) = 0$. Subtracting this term from $f(\mathbf{x}^*, \boldsymbol{\theta})$ does not change its value, that is,

$$f(\phi(\boldsymbol{\theta}), \boldsymbol{\theta}) = f(\phi(\boldsymbol{\theta}), \boldsymbol{\theta}) - \boldsymbol{\lambda}^{\mathrm{T}} g(\phi(\boldsymbol{\theta}), \boldsymbol{\theta}).$$

Differentiating both sides the above equation with respect to $\boldsymbol{\theta}$ and using the chain rule, we have

$$\nabla_{\boldsymbol{\theta}} f(\phi(\boldsymbol{\theta}), \boldsymbol{\theta}) = D\phi(\boldsymbol{\theta})^{\mathrm{T}} \nabla_{\mathbf{x}} f(\mathbf{x}^*, \boldsymbol{\theta}) + \nabla_{\boldsymbol{\theta}} f(\mathbf{x}^*, \boldsymbol{\theta}) - \left[D_{\mathbf{x}} g(\mathbf{x}^*, \boldsymbol{\theta}) D\phi(\boldsymbol{\theta}) + D_{\boldsymbol{\theta}} g(\mathbf{x}^*, \boldsymbol{\theta}) \right]^{\mathrm{T}} \boldsymbol{\lambda}$$

$$= D\phi(\boldsymbol{\theta})^{\mathrm{T}} \left[\nabla_{\mathbf{x}} f(\mathbf{x}^*, \boldsymbol{\theta}) - D_{\mathbf{x}} g(\mathbf{x}^*, \boldsymbol{\theta})^{\mathrm{T}} \boldsymbol{\lambda} \right] + \nabla_{\boldsymbol{\theta}} f(\mathbf{x}^*, \boldsymbol{\theta}) - D_{\boldsymbol{\theta}} g(\mathbf{x}^*, \boldsymbol{\theta})^{\mathrm{T}} \boldsymbol{\lambda}$$

$$= \nabla_{\boldsymbol{\theta}} f(\mathbf{x}^*, \boldsymbol{\theta}) - D_{\boldsymbol{\theta}} g(\mathbf{x}^*, \boldsymbol{\theta})^{\mathrm{T}} \boldsymbol{\lambda}$$

$$= \nabla_{\boldsymbol{\theta}} L(\mathbf{x}^*, \boldsymbol{\theta})$$

as required. In the second equality above

$$\nabla_{\mathbf{x}} f(\mathbf{x}^*, \boldsymbol{\theta}) - D_{\mathbf{x}} g(\mathbf{x}^*, \boldsymbol{\theta})^{\mathrm{T}} \boldsymbol{\lambda} = \mathbf{0}$$

by the Lagrange multiplier theorem.   □

7.3   Inequality-Constrained Optimization

In many economics applications the equality constraints are too restrictive. For example, in consumer analysis, instead of restricting a consumer to select bundles on the budget hyperplane, it may be desirable to expand the feasible set to a set in $\mathbb{R}^n$ defined by the intersection of the half-spaces $\mathbf{p}^{\mathrm{T}} \mathbf{x} \leq M, x_1 \geq 0, \ldots, x_n \geq 0$. The intersection of these $n + 1$ half-spaces is a compact and convex set called a **polytope**. Formally, an inequality-constrained optimization problem can be stated as

$$\max_{\mathbf{x}} \quad f(\mathbf{x})$$

$$\text{subject to} \quad g(\mathbf{x}) \leq \mathbf{0},$$

where $f : \mathbb{R}^n \to \mathbb{R}$ and $g : \mathbb{R}^n \to \mathbb{R}^m$ are $\mathcal{C}^1$ functions. Notice that there is no limitation on m as long as the feasible set $G = \{\mathbf{x} : g(\mathbf{x}) \leq \mathbf{0}\}$ is not empty.

Suppose $\mathbf{x}^* \in G$ is a local maximum of f and $M = \{1, \ldots, m\}$. We can classify the components of g, which can be written as $g_1(\mathbf{x}) \leq 0, \ldots, g_m(\mathbf{x}) \leq 0$ into two groups. The first group $B(\mathbf{x}^*) \subseteq M$, called the set of **binding constraints**, is defined by those components that the constraints at $\mathbf{x}^*$ are equalities, that is,

$$B(\mathbf{x}^*) = \{i : g_i(\mathbf{x}^*) = 0\}.$$

The second group, called the set of **slack constraints**, is defined by those components of g at $\mathbf{x}^*$ with strict inequalities,

$$S(\mathbf{x}^*) = \{j : g_j(\mathbf{x}^*) < 0\}.$$

For all $j \in S(\mathbf{x}^*)$, the constraint $g_j(\mathbf{x}^*) < 0$ is non-binding. It means that there exists a neighbourhood of $\mathbf{x}^*$ contained by the intersection of these constraints, which has no effect on the maximum status of $\mathbf{x}^*$ even if the constraints are relaxed. On the other hand, for all $i \in B(\mathbf{x}^*)$, $\mathbf{x}^*$ is the solution to the problem

$$\max_{\mathbf{x}} \quad f(\mathbf{x})$$

$$\text{subject to} \quad g_i(\mathbf{x}) = 0, \quad i \in B(\mathbf{x}^*). \tag{7.5}$$

By the Lagrange Multiplier Theorem, there exists λ_i's such that

$$\nabla f(\mathbf{x}^*) = \sum_{i \in B(\mathbf{x}^*)} \lambda_i \nabla g_i(\mathbf{x}^*). \tag{7.6}$$

The signs of the λ_i's can be asserted by considering the following maximization problem:

$$\max_{\mathbf{x}} \quad f(\mathbf{x})$$

$$\text{subject to} \quad g_i(\mathbf{x}) = \theta_i, \quad i \in B(\mathbf{x}^*). \tag{7.7}$$

Each of the binding equation in (7.5) is relaxed by replacing the zero by the parameter θ_i in (7.7). Using the envelope theorem, for each $i \in B(\mathbf{x}^*)$,

$$\frac{\partial f(\phi(\theta), \theta)}{\partial \theta_i} = \lambda_i.$$

Since relaxing the constraints will only increase the value of $f(\phi(\theta), \theta)$, $\lambda_i \geq 0$ for all $i \in B(\mathbf{x}^*)$. Since $g_i(\mathbf{x}^*) = 0$, we have

$$\lambda_i g_i(\mathbf{x}^*) = 0, \quad i \in B(\mathbf{x}^*).$$

Now for all $j \in S(\mathbf{x}^*)$, we can set $\lambda_j = 0$ so that

$$\lambda_j g_j(\mathbf{x}^*) = 0, \quad j \in S(\mathbf{x}^*)$$

as well. With $\lambda_j = 0$ for all $j \in S(\mathbf{x}^*)$, (7.6) can be written as

$$\nabla f(\mathbf{x}^*) = \sum_{i \in B(\mathbf{x}^*)} \lambda_i \nabla g_i(\mathbf{x}^*) + \sum_{j \in S(\mathbf{x}^*)} \lambda_j \nabla g_j(\mathbf{x}^*)$$

$$= \sum_{i=1}^{m} \lambda_i \nabla g_i(\mathbf{x}^*)$$

$$= Dg(\mathbf{x}^*)^{\mathsf{T}} \lambda,$$

where $\lambda \in \mathbb{R}^m$. The signs of the λ_i's and $g_i(\mathbf{x}^*)$'s can be summarized by the complementary slackness condition: For $i = 1, \ldots, m$,

$$\lambda_i \geq 0, \quad g_i(\mathbf{x}^*) \leq 0, \quad \lambda_i g_i(\mathbf{x}^*) = 0.$$

We have shown the following result.

Theorem 7.5 (Kuhn-Tucker Theorem) *Let $f : \mathbb{R}^n \to \mathbb{R}$ and $g : \mathbb{R}^n \to \mathbb{R}^m$ be C^1 functions. Suppose that $\mathbf{x}^*$ is a local maximum of f on the set $G = \{g(\mathbf{x}) \leq \mathbf{0}\}$. Then there exists a unique vector $\lambda \in \mathbb{R}^m$ such that*

$$\nabla f(\mathbf{x}^*) = Dg(\mathbf{x}^*)^{\mathrm{T}}\lambda;$$

$$\lambda \geq \mathbf{0}; \quad and \quad g(\mathbf{x}^*)^{\mathrm{T}}\lambda = 0.$$

In practice we construct the Lagrangian function

$$L(\mathbf{x}, \lambda) = f(\mathbf{x}) - \sum_{i=1}^{m} \lambda_i g_i(\mathbf{x})$$

as before but we set the gradient of L with respect to $\mathbf{x}$ only to zero and not λ. Instead we impose the complementary slackness conditions. Therefore the necessary conditions for inequality constrained maximization are

$$\nabla_{\mathbf{x}} L(\mathbf{x}^*, \lambda) = \mathbf{0}; \tag{7.8}$$

$$\lambda \geq \mathbf{0}; \quad g(\mathbf{x}^*) \leq \mathbf{0}; \quad g(\mathbf{x}^*)^{\mathrm{T}}\lambda = 0. \tag{7.9}$$

Example 7.4 Consider the inequality-constrained optimization problem given by

$$\max_{x,y} \quad x^2 - y^2$$

$$\text{subject to} \quad x^2 + y^2 \leq 9,$$

$$y \leq 0,$$

$$y \leq x.$$

The objective function is the saddle depicted in Fig. 7.1. The first constraint is a circular disk centred at the origin with radius equal to $\sqrt{9} = 3$. The second constraint restricts the feasible set $G(x, y)$ below the x-axis, while the last constraint limits $G(x, y)$ to the area above the straight line $y = x$. We have $n = 2$, $m = 3$ and

$$g(x, y) = (x^2 + y^2 - 9, y, y - x).$$

The Lagrangian function is

$$L = x^2 - y^2 - \lambda_1(x^2 + y^2 - 9) - \lambda_2 y - \lambda_3(y - x).$$

The necessary conditions are

$$\partial L/\partial x = 2x - 2\lambda_1 x + \lambda_3 = 0, \tag{7.10}$$

$$\partial L/\partial y = -2y - 2\lambda_1 y - \lambda_2 - \lambda_3 = 0. \tag{7.11}$$

The complementary slackness conditions are

$$\lambda_1 \geq 0, \ x^2 + y^2 \leq 9, \ \lambda_1(x^2 + y^2 - 9) = 0; \tag{7.12}$$

$$\lambda_2 \geq 0, \ y \leq 0, \ \lambda_2 y = 0; \tag{7.13}$$

$$\lambda_3 \geq 0, \ y \leq x, \ \lambda_3(y - x) = 0. \tag{7.14}$$

We solve the above system by trials and errors. In the first trial, we suppose that $x^2 + y^2 = 9$. Then $\lambda_1 \geq 0$. Also, assume that $y < 0$ and $y < x$ so that $\lambda_2 = \lambda_3 = 0$. Equation (7.11) implies that $y + \lambda_1 y = 0$ so that $y = 0$ or $\lambda_1 = -1$. These contradict the assumptions that $y < 0$ and $\lambda_1 \geq 0$. Therefore a solution does not exist in this region.

In the second trial, we assume that $x^2 + y^2 = 9$, $y < x$ and $y = 0$. This gives $x = 3$ and $\lambda_3 = 0$. Equation (7.10) implies that $\lambda_1 = 1$ and Eq. (7.11) implies that $\lambda_2 = 0$. In this case the solution

$$(x^*, y^*, \lambda_1, \lambda_2, \lambda_3) = (3, 0, 1, 0, 0)$$

satisfies all the above necessary conditions.

Example 7.5 A consumer's preference relation is represented by the quasi-linear utility function

$$U(x, y) = 2x + \log(y + 4).$$

The consumer has income equal to 5 and the market prices of the two goods are $p_x = 2$ and $p_y = 1$. The utility maximization problem is

$$\max_{x,y} \quad 2x + \log(y + 4)$$

$$\text{subject to} \quad 2x + y \leq 5,$$

$$x \geq 0,$$

$$y \geq 0.$$

Again we have $n = 2, m = 3$ and

$$g(x, y) = (2x + y - 5, -x, -y).$$

The Lagrangian function is

$$L(x, y, \lambda_1, \lambda_2, \lambda_3) = 2x + \log(y + 4) - \lambda_1(2x + y - 5) + \lambda_2 x + \lambda_3 y.$$

The necessary conditions in (7.8) and (7.9) are

$$\partial L/\partial x = 2 - 2\lambda_1 + \lambda_2 = 0, \tag{7.15}$$

$$\partial L/\partial y = 1/(y+4) - \lambda_1 + \lambda_3 = 0, \tag{7.16}$$

$$\lambda_1 \geq 0, \quad 2x + y \leq 5, \quad \lambda_1(2x + y - 5) = 0, \tag{7.17}$$

$$\lambda_2 \geq 0, \quad x \geq 0, \quad \lambda_2 x = 0 \tag{7.18}$$

$$\lambda_3 \geq 0, \quad y \geq 0, \quad \lambda_3 y = 0 \tag{7.19}$$

We also solve the system by trials and errors:

1. Assume that we have an interior solution, that is, $x > 0$ and $y > 0$. Then by conditions (7.18) and (7.19), $\lambda_2 = \lambda_3 = 0$.
2. Since the utility function U is increasing in x and y, we assume that in condition (7.17) the budget constraint is binding, that is, $2x + y = 5$ and $\lambda_1 \geq 0$. Then Eqs. (7.15) and (7.16) imply that

$$\lambda_1 = 1, \quad \text{and} \quad \lambda_1 = 1/(y+4),$$

which gives $y = -3$, which contradicts the assumption that $y > 0$.
3. The above result gives us a hint that we should assume $x > 0$ and $y = 0$. Then $\lambda_2 = 0$ and $\lambda_3 \geq 0$. We also keep the assumption that $2x + y = 5$ and $\lambda_1 \geq 0$. With Eqs. (7.15) and (7.16) we have

$$2x = 5,$$

$$\lambda_1 = 1,$$

$$1/4 - \lambda_1 + \lambda_3 = 0.$$

These give $x = 5/2$ and $\lambda_3 = 3/4$.

Putting the results together, the solution

$$(x^*, y^*, \lambda_1, \lambda_2, \lambda_3) = (5/2, 0, 1, 0, 3/4)$$

satisfies the implications of the Kuhn-Tucker Theorem, with $U(x^*, y^*) = 5 + \log 4$. As an exercise you should try the assumptions that $x = 0$ and $y > 0$.

7.4 Exercises

1. Let $f : S \rightarrow \mathbb{R}$ where $S \subseteq \mathbb{R}^n$ is a convex set. Suppose f is a C^1 concave function and $\nabla f(\mathbf{x}^*) = \mathbf{0}$. Show that $\mathbf{x}^* \in S$ is a global maximum for f over S.
2. Consider the functional

$$f(x, y) = xy + 3y - x^2 - y^2.$$

(a) Find the stationary point(s) of f.

(b) Is it a maximum, minimum or neither? Explain.

3. Find the stationary point(s) of

$$f(x, y) = 1 - (x - 1)^2 - (y - 1)^2.$$

Determine whether the point(s) is (are) maximum, minimum or neither.

4. Solve the following optimization problems. Check second order conditions.

(a) $\max_{x_1, x_2} -2x_1^2 + x_1 x_2 - 2x_2^2 - 3x_1 - 3x_2$,

(b) $\min_{x_1, x_2, x_3} x_1^2 + x_2^2 + x_3^2$.

5. Find and classify the critical points of the function

$$f(x, y) = x \sin y.$$

6. A swimming pool is to be constructed with reinforced concrete. The pool will have a square base with length and width x and height h. The cost of the concrete, including labour and materials, forming the base is p per square metre. The concrete for the side walls costs q per square metre. The budget for the construction cost of the pool is M.

(a) Setting up a constrained-optimization problem to maximize the volume of the pool.

(b) Find the Lagrangian function.

(c) Find the effect on the optimized volume with respect to a change in the cost of the base material p.

7. Suppose that a function $f : \mathbb{R}^2 \to \mathbb{R}$ is given by

$$f(\mathbf{x}) = 4x_1 x_2 - x_1^2 - x_2^2.$$

(a) Find the gradient and the Hessian of f.

(b) Find a stationary point of f.

(c) Is the stationary above a maximum or a minimum? Explain.

8. Consider the unconstrained maximization of the objective function $f(\mathbf{x}, \boldsymbol{\theta})$ where $\mathbf{x} \in \mathbb{R}^n$ is the control variable and $\boldsymbol{\theta} \in \mathbb{R}^l$ is the parameter vector. The solution can be expressed as $\mathbf{x}^* = \phi(\boldsymbol{\theta})$. Find the impact of a change in $\boldsymbol{\theta}$ on $\mathbf{x}^*$, that is, find $D\phi(\boldsymbol{\theta})$. (Hint: Apply the implicit function theorem to the first-order condition.)

9. Provide an alternative proof of the Cauchy-Schwarz Inequality: Let $\mathbf{x}, \mathbf{y} \in \mathbb{R}^n$. Then

$$(\mathbf{x}^T \mathbf{y})^2 \leq (\mathbf{x}^T \mathbf{x})(\mathbf{y}^T \mathbf{y}) \qquad (7.20)$$

Hint: If either $\mathbf{x}$ or $\mathbf{y} = \mathbf{0}$, then (7.20) is trivially true. Assume $\mathbf{x}$ and $\mathbf{y} \neq \mathbf{0}$, and therefore $(\mathbf{x}^T \mathbf{x}) > 0$ and $(\mathbf{y}^T \mathbf{y}) > 0$. Now for every real number α, define $\mathbf{z} = \mathbf{x} + \alpha \mathbf{y}$ so that $\mathbf{z} \in \mathbb{R}^n$. We have

$$0 \leq \mathbf{z}^T \mathbf{z}$$
$$= (\mathbf{x} + \alpha \mathbf{y})^T (\mathbf{x} + \alpha \mathbf{y})$$
$$= \alpha^2 \mathbf{y}^T \mathbf{y} + 2\alpha \mathbf{x}^T \mathbf{y} + \mathbf{x}^T \mathbf{x}$$
$$= f(\alpha). \tag{7.21}$$

The inequality in (7.21) is true for every real number α. Now minimize $f(\alpha)$ with respect to α. Denote the minimum point as α^*. Calculate $f(\alpha^*)$ and it will turn out that the inequality $f(\alpha^*) \geq 0$ is equivalent to (7.20).

10. Suppose $y \in \mathbb{R}$ has a linear relation with $\mathbf{x} \in \mathbb{R}^k$, that is,

$$y = \mathbf{x}^T \boldsymbol{\beta},$$

where $\boldsymbol{\beta} \in \mathbb{R}^k$ is the unknown matrix representation of a linear functional. To estimate the value of $\boldsymbol{\beta}$ we carry out $n(> k)$ random experiments with different values of $\mathbf{x}$ and observe the corresponding values of y. The result can be presented in the following matrix form:

$$\mathbf{y} = X\boldsymbol{\beta} + \boldsymbol{\epsilon}$$

where $\mathbf{y} \in \mathbb{R}^n$, X is a $n \times k$ matrix, and $\boldsymbol{\epsilon} \in \mathbb{R}^n$ is the vector of random variables caused by measurement errors. The error term of each observation ϵ_i is assumed to have the same distribution with zero mean and variance σ^2. Also, assume that $E(\epsilon_i \epsilon_j) = 0$ for $i \neq j$. One way to estimate $\boldsymbol{\beta}$ is to minimize the sum of squared errors, that is, we want to minimize $\boldsymbol{\epsilon}^T \boldsymbol{\epsilon}$.
   (a) Derive an expression for $\boldsymbol{\epsilon}^T \boldsymbol{\epsilon}$.
   (b) Find the stationary point of $\boldsymbol{\epsilon}^T \boldsymbol{\epsilon}$ with respect to $\boldsymbol{\beta}$. Assuming X have rank k, derive the so-called least square estimator for $\boldsymbol{\beta}$.
   (c) Find the Hessian of $\boldsymbol{\epsilon}^T \boldsymbol{\epsilon}$. Show that it is positive definite so that you have indeed found a minimum point.

11. Suppose that

$$f(x) = \frac{1}{4}x^4 + \frac{1}{2}ax^2 + bx,$$

where a and b are parameters.
   (a) Find the set of critical points of f.
   (b) Apply the implicit function theorem to the necessary condition in part (a) to find the rate of change of the critical point x with respect to a and b.
   (c) Find the set of points (a, b) such that the implicit function theorem fails to apply.
The set in part (a) is called a cusp manifold (Fig. 7.2). The set in part (c) is called the bifurcation set. The function f is useful in the study of stability of dynamical systems. See, for example, Poston and Stewart (1978, p. 78–83) for details.

Fig. 7.2 A cusp manifold

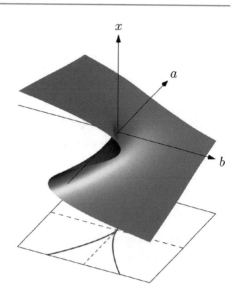

12. Verify that the other two points in Table 7.1 are minimum points.
13. In the constrained optimization problem

$$\max_{\mathbf{x} \in G(\boldsymbol{\theta})} f(\mathbf{x}, \boldsymbol{\theta}),$$

suppose that f is strictly quasi-concave in $\mathbf{x}$ and $G(\boldsymbol{\theta})$ is convex. Prove that a local maximum is a strict global maximum.

14. Given the production function $f(\mathbf{x}) = (x_1 x_2)^{1/2}$ where x_1, x_2 are input quantities, a firm is trying to minimize the input cost given a particular output level y. Suppose the market prices for x_1 and x_2 are \$2 and \$5 per unit respectively.
    (a) Set up the optimization problem.
    (b) Find the response of the cost with respect to an increase in output.

15. Find the bordered Hessian of the consumer maximization problem in Example 7.3.

16. Let $f : \mathbb{R}_+^n \to \mathbb{R}_+$ be the C^2 production function of a firm. For $y \in f(\mathbb{R}_+^n)$ and $\mathbf{p} \in \mathbb{R}_{++}^n$, define the producer's cost function by

$$C(\mathbf{p}, y) = \min_{\mathbf{x}}\{\mathbf{p}^{\mathsf{T}}\mathbf{x} : f(\mathbf{x}) = y\}.$$

Prove that C satisfies the following three properties:
    (a) $C(\alpha\mathbf{p}, y) = \alpha C(\mathbf{p}, y)$ for $\alpha > 0$;
    (b) $\mathbf{p}^1 \le \mathbf{p}^2$ implies $C(\mathbf{p}^1, y) \le C(\mathbf{p}^2, y)$; and
    (c) $C(\mathbf{p}, y)$ is concave in $\mathbf{p}$ for fixed y.

17. Consider the cost minimization problem

$$\min_{\mathbf{z}} \quad w_1 z_1 + w_2 z_2$$

subject to $\quad y = z_1 z_2.$

where $\mathbf{z} \in \mathbb{R}_+^2$ and $\mathbf{w} \in \mathbb{R}_{++}^2$. Show that the solution is $f(z_1^*, z_2^*) = 2(w_1 w_2 y)^{1/2}$ by representing it as an unconstrained minimization problem in one variable.

18. Find the maxima and minima of the function

$$f(x, y) = 1/x + 1/y$$

subject to the constraint $1/x^2 + 1/y^2 = 1/4$.

19. Solve the following constrained maximization problem:

$$\max_{\mathbf{z}} \quad a_0 - \mathbf{a}^\mathsf{T} \mathbf{z} + \frac{1}{2} \mathbf{z}^\mathsf{T} A \mathbf{z}$$

subject to $D^\mathsf{T} \mathbf{z} = \mathbf{b}$ where D is a $n \times m$ matrix of rank m, $A = A^\mathsf{T}$ is a negative definite $n \times n$ matrix, and $m < n$. Check whether your solution satisfies a set of second-order sufficient conditions.

20. Suppose a consumer maximizes her utility by buying two products with amounts x_1 and x_2 according to the utility function $U(x_1, x_2) = x_1^\alpha x_2^{1-\alpha}$. Her budget constraint is $p_1 x_1 + p_2 x_2 = y$ where p_1, p_2 are the prices and y is her weekly expenditure on the two products.

    (a) Find the weekly ordinary demand function for the two products.

    (b) Define the indirect utility function $V(p_1, p_2, y) = U(x_1^*, x_2^*)$ to be the value function of the optimal choice. Show that

$$\frac{\partial V(p_1, p_2, y)}{\partial y} = \lambda,$$

    where λ is the Lagrange multiplier.

21. Find the maximum of the function $f(x, y) = x^2 - y^2$ in the constraint set $\{(x, y) \in \mathbb{R}^2 : x^2 + y^2 \le 1\}$.

22. Solve the following maximization problem:

$$\max_{x, y} \quad 2 \log x + y$$

subject to $\quad x + 2y \le 1, x \ge 0, y \ge 0$.

23. The utility function of a consumer is given by $U(x_1, x_2) = (x_1 x_2)^{1/2}$, where prices for goods 1 and 2 are $p_1 > 0$ and $p_2 > 0$. Given that the consumer has

income $y > 0$, solve the utility maximization problem subject to the constraints, $p_1 x_1 + p_2 x_2 \le y$, $x_1 \ge 0$, $x_2 \ge 0$ as follows:

(a) Show that this can be reduced to an equality-constrained problem.

(b) Find the critical point(s) of the Lagrangian.

(c) Check the second-order conditions.

24. Suppose that a consumer's preferences are represented by an increasing and quasi-concave utility function C^2 function $u = f(\mathbf{x})$, where $\mathbf{x} \in \mathbb{R}_+^n$ is the consumption bundle. The budget constraint is given by $\mathbf{p}^T \mathbf{x} = y$, where $\mathbf{p} \in \mathbb{R}_{++}^n$ is the price vector and $y > 0$ income.

(a) Set up the utility maximization problem as an equality-constrained optimization problem.

(b) Let $\mathbf{x}^*$ be the optimal consumption bundle and $V(\mathbf{p}, y)$ be the value function (called the indirect utility function). Use the envelope theorem to prove Roy's identity:

$$x_i^* = -\frac{\partial V(\mathbf{p}, y)/\partial p_i}{\partial V(\mathbf{p}, y)/\partial y}, \quad i = 1, \ldots, n$$

25. A competitive firm produces a single output Q with capital K and labour L with production technology according to an increasing and strictly concave production function

$$Q = F(L, K).$$

The market price of the output is p and the market prices of capital and labour are r and w respectively.

(a) Set up the profit maximization problem as an equality-constrained maximum problem. Identify the control variables.

(b) Set up the Lagrangian.

(c) The value function, called the profit function, is written as $\pi(p, w, r)$. Prove the Hotelling's lemma:

$$\frac{\partial \pi(p, w, r)}{\partial p} = Q^*,$$

where Q^* denotes the optimal level of output.

26. Use the Kuhn-Tucker Theorem to solve the following optimization problem:

$$\max_{x, y} \quad [-(x - 4)^2 - (y - 4)^2]$$

$$\text{subject to} \quad x + y \le 4,$$

$$x + 3y \le 9.$$

27. A consumer who buys two goods has a utility function $U(x_1, x_2) =$ $\min\{x_1, x_2\}$. Given income $y > 0$ and prices $p_1 > 0$ and $p_2 > 0$.
   (a) Describe the consumer's utility maximization problem.
   (b) Does the Weierstrass theorem apply to this problem?
   (c) Can the Kuhn-Tucker theorem be used to obtain a solution?
   (d) Solve the maximization problem.
   Explain your answers.

28. Solve the following maximization problem:

$$\text{Maximize} \quad xy$$
$$\text{subject to } x + 2y \leq 5, \ x \geq 0, \ y \geq 0.$$

29. Solve the following maximization problem:

$$\text{Maximize } x + \log(1 + y)$$
$$\text{subject to} \quad x + 2y = 1, \ x \geq 0, \ y \geq 0.$$

30. An inequality-constrained optimization problem is given by

$$\max_{x, y} \quad x^2 - y^2$$
$$\text{subject to} \quad x^2 + y^2 \leq 4,$$
$$y \geq 0,$$
$$y \leq x.$$

   (a) Derive the Kuhn-Tucker conditions by assuming that only the first constraint is binding. Determine whether a solution exists with this assumption.
   (b) Find the solution of the problem.

References

Debreu, G. (1952). Definite and semidefinite quadratic forms. *Econometrica, 2*(20), 295–300.
Poston, T., & Stewart, I. (1978). *Catastrophe theory and its applications*. London: Pitman.

Probability

8

Most graduate level econometrics textbooks require the readers to have some background in intermediate level mathematical probability and statistics. A lot of students, however, took business and economic statistics and then econometric courses in the undergraduate program. This chapter gives a quick introduction to the formal theory. It is by no means a replacement for the proper training. Knowledge in probability theory is also essential in the studies of economics of risk and uncertainty, financial economics, game theory and macroeconomics.

8.1   Basic Definitions

Throughout the history people involved in decision making and gambling activities have developed an intuitive idea of probability, which is the likeliness of some events would happen. The formal theory was first constructed by Blaise Pascal and Pierre de Fermat in the seventeenth century (see Ross 2004).

We start with the idea of a **random experiment** or **random trial**. Before the experiment, it is assumed that we know the set of all possible **outcomes**, which is called the **sample space**, denoted by Ω. At the end of the experiment, one outcome is realized.

Example 8.1 A coin is tossed once, there are two possible outcomes:

© Springer Nature Switzerland AG 2019
K. Yu, *Mathematical Economics*, Springer Texts in Business and Economics,
https://doi.org/10.1007/978-3-030-27289-0_8

Let us call the outcome on the left "head", or simply H, and the one on right "tail", or T. Then the random experiment is the tossing of the coin, and the sample space is $\Omega = \{H, T\}$. At the end of the experiment either H or T is the outcome, but not both.

Example 8.2 A die is thrown once. The sample space Ω consists of six outcomes:

Often we want to represent the outcomes by numbers. So we define a function $X : \Omega \rightarrow \mathbb{R}$ called a **random variable**. Random variables are usually denoted by capital letters, say, X. In Example 8.1, we can define

$$X(H) = 0, \quad X(T) = 1,$$

so that the range of the random variable X is $\{0, 1\}$. In Example 8.2, the natural way to assign values to the variable is by the number of dots on the face of the die. In this case the range of X is $\{1, 2, 3, 4, 5, 6\}$. It is clear that a random variable is everywhere defined but not necessary one-to-one or onto.

Nevertheless, the assignment of a random variable to the sample space may not be meaningful, as the following example shows.

Example 8.3 A card is chosen randomly from a deck of cards. The sample space is

$$\Omega = \{\spadesuit A, \spadesuit 2, \ldots, \spadesuit 10, \spadesuit J, \spadesuit Q, \spadesuit K, \heartsuit A, \ldots, \heartsuit K, \diamondsuit A, \ldots, \diamondsuit K, \clubsuit A, \ldots, \clubsuit K\},$$

with 52 possible outcomes. The sample space can also be seen as a product set $\Omega = S \times T$, where

$$S = \{\spadesuit, \heartsuit, \diamondsuit, \clubsuit\},$$
$$T = \{A, 2, \ldots, 10, J, Q, K\}.$$

We shall see later in Example 8.11 that independent events can be constructed from the product sets.

A subset of the sample space is called an **event**. Since events are sets, the set operations defined in Chap. 2 can be used to defined new events. For example, the event $A \cup B$ means that the outcome belongs to A or to B. Two events A and B are said to be **mutually exclusive** if A and B are disjoint. The complement of A, A^c is the event that A does not happen.

Example 8.4 Let A be the event that the randomly chosen card be a seven. Then $A = \{\spadesuit 7, \heartsuit 7, \diamondsuit 7, \clubsuit 7\}$. Let B be the event that the card is a heart, that is, $B = \{\heartsuit A, \ldots, \heartsuit K\}$. Then the event $A \setminus B$ is the event of a seven card but not $\heartsuit 7$. That is,

$$A \setminus B = \{\spadesuit 7, \diamondsuit 7, \clubsuit 7\}.$$

Example 8.5 In Example 8.1 a coin is tossed once, with $\Omega = \{H, T\}$. Suppose now that the coin is tossed twice. The sample space is

$$\Omega^2 = \{(H, H), (H, T), (T, H), (T, T)\}.$$

Let A be the event that we have two heads, B means two tails and C means one head and one tail. Then A, B, C is a partition of the sample space Ω. In other words, the events are pairwise mutually exclusive and $A \cup B \cup C = \Omega$.

The likeliness that an event will happen is represented by a number between 0 and 1 assigned to the event. Formally, a **probability measure** is a function P which maps the set of all events of a sample space Ω into $[0, 1]$ which satisfies the following axioms:

1. $P(A) \geq 0$ for every $A \subseteq \Omega$.
2. $P(\Omega) = 1$.
3. If A and B are mutually exclusive, then $P(A \cup B) = P(A) + P(B)$.

In assigning probabilities to events, we often take the frequentist approach or the subjective approach. In the first approach, we imagine that we can repeat the same random experiment many times, say N and observe the number of times n that an event A occurs. Then we assume that $P(A) = n/N$ as $N \to \infty$. In the subjective approach, the analyst assigns the probability according to her subjective opinion or based on some deductive reasonings. In simple cases the two approaches can give the same result. For example, if the sample space Ω is finite, and each outcome is considered equally likely to occur, then the probability of an event A can be defined as

$$P(A) = \frac{|A|}{|\Omega|}. \tag{8.1}$$

That is, the likeliness of event A happens is the ratio of the number of outcomes in A and the size of the sample space.

Example 8.6 Suppose a coin is tossed once. If we assume that the coin is fair, there is no reason to believe that the outcome H is more likely than T, or vice versa. Therefore it is natural to assign the probabilities

$$P(\{H\}) = P(\{T\}) = 1/2.$$

Example 8.7 Consider the experiment in Example 8.5. Again if we assume a fair coin, the probabilities of the four outcomes are the same. Then $P(A) = P(B) = 1/4$. By the axioms above $P(C) = 1/2$.

Example 8.8 It is often more convenient to define the probabilities on the random variable instead of events. Consider the experiment in throwing a fair die in Example 8.2. It is natural to assign equal probability of $1/6$ to each outcome in Ω. Therefore

$$P(X = 1) = P(X = 2) = \cdots = P(X = 6) = 1/6.$$

The probability of the event that X is less than or equal to 2 can be written as

$$P(X \leq 2) = P(X = 1) + P(X = 2) = 1/3.$$

Similarly,

$$P(X \text{ is odd}) = P(X = 1) + P(X = 3) + P(X = 5) = 1/2.$$

The following results can be proven directly from the three axioms of probability and left as an exercise.

Theorem 8.1 *Let A and B be events of a sample space Ω. Then*

1. $P(A^c) = 1 - P(A)$.
2. *If $A \subseteq B$, then $P(A) \leq P(B)$.*
3. $P(A \cup B) = P(A) + P(B) - P(A \cap B)$.
4. $P(A \setminus B) = P(A) - P(A \cap B)$.
5. $P(\varnothing) = 0$.

When the sample space Ω is a finite or countable set, the set of all events is the power set of Ω. The probability function defined above is well defined. Things are a bit more complicated if Ω is uncountable. In order for the probability function to be well defined, we often narrow the set of events to a collection of sets called a σ-algebra or σ-field, which is closed under countable unions, intersections and complements. Formally, a σ-**field** $\mathcal{F}$ is a collection of subsets of the sample space Ω which satisfies the following three axioms:

1. $\varnothing \in \mathcal{F}$,
2. For every $A \in \mathcal{F}$, the complement, $A^c \in \mathcal{F}$,
3. If $\{A, B, C, \dots\}$ are in $\mathcal{F}$, then $A \cup B \cup C \cup \cdots$ is also in $\mathcal{F}$.

By axioms 1 and 2, the sample space Ω is always in $\mathcal{F}$. Also, axioms 2 and 3 and De Morgan's Law imply that any countable intersection of events in $\mathcal{F}$ is also in $\mathcal{F}$.

8.2 Conditional Probability

Suppose that we are concerned with the occurrence of a particular event A within a sample space Ω. Before the outcome is revealed, we are given an additional piece of information that event B has occurred. Will this information change our perspective on the probability of the event A? Prior to the random experiment, suppose we know that the probabilities of event A is $P(A)$. We also know $P(B)$, and $P(A \cap B)$. Now given that we know event B has been realized, is the probability of A still $P(A)$? Formally, we are evaluating the **conditional probability** of event A given that event B has occurred, which is denoted by $P(A|B)$. Consider the following two special cases:

1. Suppose that A and B are mutually exclusive. If B is realized, then A cannot happen. In this case $P(A|B) = 0$.
2. Suppose that $B \subseteq A$. By definition B is realized means that A is also realized so that $P(A|B) = 1$.

In the general case, it depends on $P(A \cap B)$. Since we know that B has occurred, the sample space is no longer Ω but B. Therefore we have to "rescale" $P(A \cap B)$ using B as the sample space so that all the mutually exclusive events still sum to 1. It follows that

$$P(A|B) = P(A \cap B) \times \frac{P(\Omega)}{P(B)} = \frac{P(A \cap B)}{P(B)}. \tag{8.2}$$

Notice that the two special cases above satisfy Eq. (8.2) as well. In the first case, $P(A \cap B) = 0$ and in the second $P(A \cap B) = P(B)$. Also, since event B is realized, $P(B) \neq 0$ so that Eq. (8.2) is well defined.

Example 8.9 Consider the probability of getting the queen of heart from a random draw from a desk of cards, which is $P(A) = 1/52$. Before the card is revealed, we are told that the card is a heart. So it is clear that from the new information the probability of getting the queen of heart is $1/13$. Formally, A is the event of getting the queen of heart. B is the event of getting a heart, with $P(B) = 1/4$. In this case $A \subseteq B$ so that $P(A \cap B) = P(A) = 1/52$. It follows that

$$P(A|B) = \frac{P(A \cap B)}{P(B)} = \frac{1/52}{1/4} = \frac{1}{13}.$$

Psychologists Amos Tversky and Daniel Kahneman (1974) point out that intuition is very unreliable when people make subjective probabilities. The following example, taken from Economist (1999), illustrates the importance of the concept of conditional probability. Most people, unfortunately probably including your family doctor, get the answer wrong the first time.

Example 8.10 You are given the following information. (i) In random testing, you test positive for a disease. (ii) In 5% of cases, this test shows positive even when the subject does not have the disease. (iii) In the population at large, one person in 1,000 has the disease. What is the probability that you have the disease?

Let A be the event that you have the disease, and B be the event that you test positive for the disease. So $P(A) = 0.001$ and $P(B) = 0.05$. We can assume that if you have the disease you will be tested positive as well. Therefore, $A \subseteq B$ so that $P(A \cap B) = P(A) = 0.001$. It follows that

$$P(A|B) = \frac{P(A \cap B)}{P(B)} = \frac{0.001}{0.05} = 0.02.$$

That is, even though you test positive, the probability of having the disease is only 2%. As the article points out, most people's answers is 95%. They fail to include the information in (iii) in their subjective probabilities.

Suppose that A and B are two events from a sample space Ω. We say that A is **independent** of B if $P(A|B) = P(A)$. That is, the probability of event A is unaffected by whether event B happens or not. Using Eq. (8.2), we have

$$P(A|B) = \frac{P(A \cap B)}{P(B)} = P(A),$$

which gives

$$P(A \cap B) = P(A)P(B). \tag{8.3}$$

It can be shown easily that if A is independent of B, then B is also independent of A so that the relation is symmetric. Equation (8.3) says that if two events are independent, then the probability that they both occur is the product of their probabilities.

Example 8.11 Let the event A be drawing a 7 from a deck of cards and B be drawing a diamond. Then $P(A) = 1/13, P(B) = 1/4$ and $P(A \cap B) = 1/52$. It follows that

$$P(A|B) = \frac{P(A \cap B)}{P(B)} = \frac{1/52}{1/4} = \frac{1}{13} = P(A),$$

$$P(B|A) = \frac{P(A \cap B)}{P(A)} = \frac{1/52}{1/13} = \frac{1}{4} = P(B),$$

so that A and B are independent.

Independent events arise from situations when the sample spaces are product sets, as Examples 8.3 and 8.11 show. A common situation is when we repeat a random experiment n times, each with a sample space Ω. The sample space of all the n trials is Ω^n. Then any event in one particular trial is independent of any event in another trial. For example, if we roll a die ten times, even the outcomes of the first nine are all even, the probability of the last roll being odd is still one half.

By the definition of conditional probability, we have

$$P(B|A) = \frac{P(A \cap B)}{P(A)},$$

which gives

$$P(A \cap B) = P(B|A)P(A). \tag{8.4}$$

Substitute Eq. (8.4) into (8.2), we obtain **Bayes' Theorem**:

$$P(A|B) = \frac{P(B|A)P(A)}{P(B)}.$$

There are many interpretations of Bayes' Theorem. Here we provide one in line with the question we ask at the beginning of the section. Suppose we have a prior believe that the probability of event A happens is $P(A)$. Then we learn that another event B has occurred. Given the new information, how would we update our believe on event A? Bayes' Theorem says that we should update $P(A)$ by multiplying it with the factor $P(B|A)/P(B)$, the ratio of the conditional probability of B given A and the probability of B.

8.3   Probability Distributions

Consider a random experiment with a countable sample space Ω. Then the range of the random variable X is a countable set of real numbers with the same cardinality as Ω, which can be denoted by $S = \{x_1, x_2, x_3, \ldots \}$. The set S is called the **support** of the random variable X. We define the **probability mass function** $f : S \to [0, 1]$ as

$$f(x) = P(X = x).$$

That is, $f(x)$ is the probability of the random variable X equal to x.

Example 8.12 Consider the experiment of rolling a fair die once. The support of the random variable X is $S = \{1, 2, 3, 4, 5, 6\}$. The probability mass function is

$$f(x) = \begin{cases} 1/6 & x = 1, 2, \ldots, 6, \\ 0 & \text{elsewhere.} \end{cases}$$

By the axioms of probability, the probability mass function (pmf) satisfies the following two properties:

1. $f(x) \geq 0$ for all $x \in \mathbb{R}$,
2. $\sum_{x \in S} f(x) = 1$.

The **cumulative distribution function**, or simply distribution function $F : S \to [0, 1]$ is defined as

$$F(x) = P(X \leq x). \tag{8.5}$$

It is clear that F is an increasing function of x, and

$$F(x) = \sum_{X \leq x} f(X).$$

Example 8.13 The distribution function in Example 8.12 is

$$F(x) = \begin{cases} 0 & x < 1, \\ n/6 & n \leq x < n+1, \ n = 1, 2, 3, 4, 5, \\ 1 & x \geq 6. \end{cases}$$

When the sample space Ω is finite or countable, X is called a **discrete** random variable. Now we consider the case when Ω is uncountable, which is often the case when we measure a quantity such as time, price, quantity, inflation, expenditure and utility. Events are often described by intervals on the real line. In this case we call X a **continuous** random variable. The distribution function of X is the same as defined in Eq. (8.5). The distribution function of a continuous random variable is increasing and also continuous. Unlike the discrete case, there is no x such that $P(X = x)$ has a positive value, that is, $P(X = x) = 0$ for all $x \in \mathbb{R}$. In many cases, the distribution function of a continuous random variable can be expressed as

$$F(x) = \int_{-\infty}^{x} f(t) \, dt,$$

where f is called the **probability density function** (pdf) of X. By the Fundamental Theorem of Calculus,

$$F'(x) = f(x)$$

on the support of X, which is defined as

$$S = \{x \in \mathbb{R} : f(x) > 0\}.$$

From the above definition of F, the probability of an event $A = [a, b]$ is

$$P(a \leq X \leq b) = F(b) - F(a) = \int_a^b f(t)\, dt.$$

By the axioms of probability, f is required to satisfy the following properties:

1. $f(x) \geq 0$ for all $x \in \mathbb{R}$,
2. $\int_{-\infty}^{\infty} f(t)\, dt = 1$.

Example 8.14 A random experiment involves choosing a point X in the interval $[0, 1]$. (Most electronic calculators have a function generating random numbers.) The distribution function of X is

$$F(x) = \begin{cases} 0 & x < 0, \\ x & 0 \leq x \leq 1, \\ 1 & x > 1. \end{cases}$$

The pdf of X is

$$f(x) = \begin{cases} 1 & 0 \leq x \leq 1, \\ 0 & \text{elsewhere.} \end{cases}$$

The probability that the number is less than $1/2$ is

$$P(X < 1/2) = F(1/2) = \int_{-\infty}^{1/2} f(t)\, dt = \int_0^{1/2} 1\, dt = \frac{1}{2}.$$

Other similar distributions, with the pdf equal to a constant value on an interval, are generally called uniform distributions.

8.4 Mathematical Expectations

Suppose that X is a random variable with a pmf or pdf $f(x)$. Let $u(X)$ be a function of X. Suppose further that, in the discrete case,

$$\sum_{x \in S} |u(x)| f(x) \leq \infty, \tag{8.6}$$

or in the continuous case,

$$\int_{-\infty}^{\infty} |u(x)| f(x)\, dx \le \infty, \tag{8.7}$$

then we can define the **mathematical expectation** of $u(X)$ in the discrete case as

$$E[u(X)] = \sum_{x \in S} u(x) f(x),$$

and in the continuous case,

$$E[u(X)] = \int_{-\infty}^{\infty} u(x) f(x)\, dx.$$

Given a random variable X, let $V(X)$ be the set of functions of X that satisfy inequality (8.6) in the discrete case and inequality (8.7) in the continuous case. Then it can be shown that $V(X)$ is a vector space. By the properties of integration, mathematical expectation is a linear functional on $V(X)$. That is, for any $u, v \in V(X)$ and $\alpha \in \mathbb{R}$,

$$E[\alpha u(X) + v(X)] = \alpha E[u(X)] + E[v(X)]. \tag{8.8}$$

Some of the important expectations for the continuous case are discussed as follows. The counterparts for the discrete case are similar by using summation instead of integration.

If $u(X) = X$, then

$$E(X) = \int_{-\infty}^{\infty} x f(x)\, dx$$

is called the **expected value** of X or simply the **mean** of X, and sometimes denoted by μ. Intuitively, it gives the notion of the average of the outcomes if we repeat the random experiment many times. The concept is also analogous to the centre of gravity of a rigid body in physics. For this reason the mean is sometimes called the first moment of the probability distribution about the point zero.

If $u(X) = (X - \mu)^2$, then

$$E[(X - \mu)^2] = \int_{-\infty}^{\infty} (x - \mu)^2 f(x)\, dx$$

is called the **variance** of X and is denoted by σ^2 or $\text{Var}(X)$. It takes the average of the square of the distance of each of the outcomes weighted by the probabilities. Therefore σ^2 is a measure of how spread out the outcomes are away from the mean. For this reason it is sometimes called the second moment of the distribution about the mean. The positive square root of the variance, σ, is called the **standard**

deviation of X. Some statisticians prefer the use of a dimensionless quantity to measure the dispersion, which is called the **coefficient of variation** and is defined as

$$c_v = \frac{\sigma}{\mu},$$

provided that $\mu \neq 0$.

By the linear property described in Eq. (8.8), we have

$$
\begin{aligned}
E[(X - \mu)^2] &= E(X^2 - 2\mu X + \mu^2) \\
&= E(X^2) - 2\mu E(X) + \mu^2 \\
&= E(X^2) - \mu^2,
\end{aligned}
$$

which is sometimes a convenient way to calculate σ^2.

Example 8.15 Consider again the experiment of rolling a fair die once. The pmf is given in Example 8.12. The mean is

$$\mu = E(X) = \sum_{x=1}^{6} \frac{1}{6} x = 3.5.$$

The variance is

$$\sigma^2 = E[(X - \mu)^2] = \sum_{x=1}^{6} \frac{1}{6}(x - \mu)^2 = 2.917.$$

The standard derivation is $\sigma = \sqrt{2.917} = 1.708$. The coefficient of variation is $c_v = 1.708/3.5 = 0.488$.

Example 8.16 Consider a random variable X with a uniform pdf

$$
f(x) = \begin{cases} 1/(2a) & -a < x < a, \\ 0 & \text{elsewhere.} \end{cases}
$$

We have

$$\mu = E(X) = \int_{-a}^{a} \frac{1}{2a} x \, dx = 0,$$

and

$$\sigma^2 = E[(X - \mu)^2] = \int_{-a}^{a} \frac{1}{2a} x^2 \, dx = \frac{a^2}{3}.$$

This gives $\sigma = a/\sqrt{3}$. The coefficient of variation c_v is undefined since $\mu = 0$.

Example 8.17 Let us repeat Example 8.16 with the pdf

$$f(x) = \begin{cases} 1/(4a) & -2a < x < 2a, \\ 0 & \text{elsewhere.} \end{cases}$$

In this case $\mu = 0$ and $\sigma = 2a/\sqrt{3}$. The standard derivation is two times that of the previous example since the distribution is more spread out.

The following results involve simple functions of a random variable. The proofs are left as exercises.

Theorem 8.2 *Suppose that X is a random variable with mean μ and variance σ^2.*

1. Let $Y = a + bX$, where $a, b \in \mathbb{R}$ and $b \neq 0$. Then

$$E(Y) = a + b\mu, \quad \text{Var}(Y) = b^2\sigma^2.$$

2. Let

$$Z = \frac{X - \mu}{\sigma}.$$

Then Z is called the standardized random variable of X and has mean equal to 0 and variance 1.

The third moment of a standardized random variable X is called the **coefficient of skewness**, c_s, and is a measure of the symmetry of the distribution about the mean. Formally,

$$c_s = \frac{E[(X - \mu)^3]}{\sigma^3}.$$

The coefficient is negative when the distribution of X is skewed to the right, positive when skewed to the left and zero when the distribution is symmetric about the mean. It is straightforward to show that

$$c_s = \frac{E(X^3) - 3\mu\sigma^2 - \mu^3}{\sigma^3}. \tag{8.9}$$

The fourth moment of the standardized random variable is defined as the **coefficient of kurtosis**, which is

$$c_k = \frac{E[(X - \mu)^4]}{\sigma^4}.$$

There are different interpretations of c_k. The most common one is that it is a measure of the weight of the tails of the distribution. A large coefficient means that X has fat tails. The concept is similar to the moment of inertia of beams and columns in structural mechanics.[1]

8.5   Some Common Probability Distributions

8.5.1   Binomial Distribution

Suppose that a random experiment is repeated n times. In each trial the probability of an event occurring is p. Let X be the random variable of the number of trials that the event occurs. Then the support of X is $S = \{0, 1, 2, \ldots, n\}$. The probability mass function of X is the binomial distribution

$$f(x) = \binom{n}{x} p^x (1 - p)^{n-x}, \quad x = 0, 1, 2, \ldots, n,$$

where

$$\binom{n}{x} = \frac{n!}{x!(n-x)!}$$

is the number of combinations of taking x outcomes from a sample space of size n without repetitions. The binomial distribution has mean $\mu = np$ and variance $\sigma^2 = np(1 - p)$.

Example 8.18 A fair die is rolled 6 times. Let A be the event that the outcome is either 1 or 2 each time. Let X be the number of times that event A occurs. Then we have $n = 6$ and $p = 1/3$. The values of the pmf of X are listed below:

x	0	1	2	3	4	5	6
$f(x)$	0.088	0.263	0.329	0.219	0.082	0.016	0.001

The mean of the distribution is $\mu = 2$ and the variance $\sigma^2 = 4/3$. The readers can verify that $\sum_{x=0}^{6} f(x) = 1$.

8.5.2   Geometric Distribution

Suppose a random experiment is repeated until a certain event A with probability p occurs. Let X be the number of trials until A first happens. Then the support of X is $\mathbb{N}$, the set of natural numbers. The pmf of X is

$$f(x) = p(1 - p)^{x-1}, \quad x = 1, 2, 3, \ldots.$$

The distribution has mean $\mu = 1/p$ and variance $\sigma^2 = (1 - p)/p^2$.

---

[1] See, for example, Timoshenko (1940, p. 343).

Example 8.19 A fair die is rolled until the event $A = \{1, 2\}$ occurs. Let X be the number of trials that A first happens. Then $p = 1/3$. The values of the pmf of the first six trial are as follows.

x	1	2	3	4	5	6
$f(x)$	0.333	0.222	0.148	0.099	0.066	0.044
$F(x)$	0.333	0.556	0.704	0.802	0.868	0.912

The mean of the distribution is $\mu = 3$ and the variance $\sigma^2 = 6$. It is interesting to point out that although the probability of A in each of the independent trial is $1/3$, the cumulative probability of $X = 3$ is only 0.704. In fact, the cumulative distribution function of a geometric distribution is

$$F(x) = 1 - (1 - p)^x, \quad x = 1, 2, 3, \dots.$$

8.5.3 Gamma and Chi-Square Distributions

For $0 < x < \infty$, define the gamma function as[2]

$$\Gamma(x) = \int_0^\infty e^{-t} t^{x-1}\, dt.$$

Using integration by parts, it can be shown that

$$\Gamma(x + 1) = x\Gamma(x).$$

It follows that for a positive integer n,

$$\Gamma(n + 1) = n!.$$

Also,

$$\Gamma(1) = \int_0^\infty e^{-t}\, dt = 1,$$

so that we define $0! = 1$ as before. A continuous random variable X has a **gamma distribution** if the pdf has the form

$$f(x) = \frac{1}{\Gamma(\alpha)\beta^\alpha} x^{\alpha-1} e^{-x/\beta}, \quad 0 < x < \infty,$$

---

[2]See Artin (1964) for the properties of the gamma function.

where $\alpha > 0$ and $\beta > 0$ are two parameters. The mean of the distribution is $\mu = \alpha\beta$ and the variance is $\sigma^2 = \alpha\beta^2$. The gamma distribution is a very good model of income distribution because of the flexibility provided by the two parameters.

A special case of the gamma distribution is when $\alpha = k/2$ and $\beta = 2$. The pdf becomes

$$f(x) = \frac{1}{\Gamma(k/2)2^{k/2}}x^{k/2-1}e^{-x/2}, \quad 0 < x < \infty.$$

It has a special name called a **chi-square distribution** with k degrees of freedom, and is denoted by $\chi^2(k)$. The distribution has mean $\mu = k$ and variance $\sigma^2 = 2k$. The χ^2 distribution has a close relationship with the normal distribution, which we shall define next.

8.5.4 Normal Distribution

The normal distribution is the most important distribution in probability theory because of the central limit theorem. The general form of the pdf has two parameters μ and σ, and is given by

$$f(x) = \frac{1}{\sqrt{2\pi}\sigma}\exp\left[-\frac{1}{2}\left(\frac{x-\mu}{\sigma}\right)^2\right], \quad -\infty < x < \infty.$$

The normal distribution is often denoted by $N(\mu, \sigma^2)$. The mean of the distribution is μ and variance σ^2. In practice, it is often convenient to transform the distribution in standardized form, with

$$Z = \frac{X-\mu}{\sigma}.$$

By Theorem 8.2, Z has mean 0 and variance 1, with pdf

$$f(z) = \frac{1}{\sqrt{2\pi}}\exp\left(-\frac{z^2}{2}\right), \quad -\infty < z < \infty.$$

The normal distribution is symmetric about its mean and therefore the coefficient of skewness c_s is zero. Because of the shape of the graph, it is often called the bell curve (see Fig. 8.1). The following two results are very useful in statistical analysis.

Theorem 8.3 *Let X be a normally distributed random variable with mean μ and variance σ^2. Let $Z = (X - \mu)/\sigma$ be the standardized random variable of X. Then $Y = Z^2$ is $\chi^2(1)$.*

Fig. 8.1 Standard normal
distribution

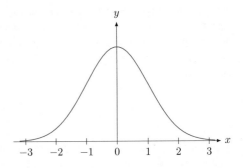

Theorem 8.4 *Let $\bar{X}_n = (X_1 + X_2 + \cdots + X_n)/n$ be the mean of n independent random samples of size n from a normal distribution that has mean μ and variance σ^2. Then $\bar{X}_n$ is $N(\mu, \sigma^2/n)$.*

The central limit theorem was presented in Chap. 3 as an example of functional convergence. We restate the theorem here in a slightly different form.

Theorem 8.5 (Central Limit Theorem) *Suppose that $\{X_1, X_2, \ldots, X_n\}$ is a sequence of outcomes from n independent random trials from a distribution with mean μ and variance σ^2. Then the random variable*

$$Z_n = \frac{\sqrt{n}(\bar{X}_n - \mu)}{\sigma},$$

where $\bar{X}_n = (X_1 + X_2 + \cdots + X_n)/n$ converges in distribution to a random variable which has a standard normal distribution $N(0, 1)$.

8.6   Multivariate Distributions

When we have two random variables, say X_1 and X_2, we consider the joint distribution with pdf $f : \mathbb{R}^2 \to \mathbb{R}_+$ where

$$\int_{-\infty}^{\infty} \int_{-\infty}^{\infty} f(x_1, x_2) \, dx_1 \, dx_2 = 1.$$

The marginal distribution of X_1 is defined by the pdf

$$f_1(x_1) = \int_{-\infty}^{\infty} f(x_1, x_2) \, dx_2.$$

The marginal distribution of X_2 can be defined similarly. The means and variances of X_1 and X_2 are then defined by their marginal pdf. Mathematical expectation is a linear operator. That is, for any scalar α,

$$E[\alpha X_1 + X_2] = \alpha E[X_1] + E[X_2].$$

In addition to their means, μ_1 and μ_2, and variances, σ_1^2 and σ_2^2, we need to ask whether they are correlated or not. The correlation is expressed by their covariance,

$$
\begin{aligned}
\sigma_{12} &= E[(X_1 - \mu_1)(X_2 - \mu_2)] \\
&= E(X_1 X_2 - \mu_2 X_1 - \mu_1 X_2 + \mu_1 \mu_2) \\
&= E(X_1 X_2) - \mu_2 E(X_1) - \mu_1 E(X_2) + \mu_1 \mu_2 \\
&= E(X_1 X_2) - \mu_1 \mu_2
\end{aligned}
$$

This is often expressed in the form

$$E(X_1 X_2) = E(X_1)E(X_2) + \mathrm{Cov}(X_1, X_2), \tag{8.10}$$

where $\mathrm{Cov}(X_1, X_2) = \sigma_{12}$.

The conditional distribution of X_2 given the value of $X_1 = x_1$ is defined as

$$f_{2|1}(x_2|x_1) = \frac{f(x_1, x_2)}{f_1(x_1)}. \tag{8.11}$$

The conditional mean or expectation of X_2 given $X_1 = x_1$ is

$$E[X_2|x_1] = \int_{-\infty}^{\infty} x_2 f_{2|1}(x_2|x_1)\, dx_2.$$

We say that the two random variables are independent if their joint pdf is separable, that is,

$$f(x_1, x_2) = f_1(x_1) f_2(x_2).$$

This implies that their covariance is zero since

$$
\begin{aligned}
\sigma_{12} &= \iint (x_1 - \mu_1)(x_2 - \mu_2) f(x_1, x_2)\, dx_1\, dx_2 \\
&= \iint (x_1 - \mu_1)(x_2 - \mu_2) f_1(x_1) f_2(x_2)\, dx_1\, dx_2 \\
&= \int (x_1 - \mu_1) f_1(x_1)\, dx_1 \int (x_2 - \mu_2) f_2(x_2)\, dx_2 \\
&= 0.
\end{aligned}
$$

It follows that (8.10) becomes

$$E(X_1 X_2) = E(X_1)E(X_2).$$

It is also clear from (8.11) that if X_1 and X_2 are statistically independent, then

$$f_{2|1}(x_2|x_1) = f_2(x_2),$$

that is, the conditional distribution of X_2 is independent of the value of X_1.

All these are conveniently expressed in matrix form as follows:

$$X = \begin{pmatrix} X_1 \\ X_2 \end{pmatrix}, \quad E(X) = \begin{pmatrix} \mu_1 \\ \mu_2 \end{pmatrix}, \quad \Sigma = \begin{pmatrix} \sigma_1^2 & \sigma_{12} \\ \sigma_{21} & \sigma_2^2 \end{pmatrix},$$

where Σ is a symmetric matrix called the variance-covariance matrix (or simply called the covariance matrix). This idea of course can be extended to a general n-vector. When X is a n-tuple of random variables, $E(X) = \mu$ is a n-vector of means and $\Sigma = E(X - \mu)(X - \mu)^T$ is a $n \times n$ symmetric covariance matrix, where the i-jth element is σ_{ij}. It can be readily shown that Σ is positive definite.

Also, if $\mathbf{c} \in \mathbb{R}^n$ is a constant vector, then

$$E(\mathbf{c}^T X) = \mathbf{c}^T E(X) = \mathbf{c}^T \mu$$

and

$$\mathrm{Var}(\mathbf{c}^T X) = \mathbf{c}^T \Sigma \mathbf{c}.$$

The most frequently joint distribution we encounter is the multivariate normal distribution, with joint pdf

$$f(\mathbf{x}) = \frac{1}{(2\pi)^{n/2}|\Sigma|^{1/2}} \exp\left\{ -\frac{1}{2}(\mathbf{x} - \mu)^T \Sigma^{-1}(\mathbf{x} - \mu) \right\}.$$

The following fact is often used in econometric analysis: If X is a random n-vector which has a multivariate normal distribution, with mean μ and covariance matrix Σ, then the statistic $(X - \mu)^T \Sigma^{-1}(X - \mu)$ has a χ^2 distribution with degree n.

In stochastic economic models we frequently employ the assumption of certainty equivalence. That is, we use the conditional mean of a random variable in an optimization problem. Care should be taken in applying the statistical rules. For example, expected utility is different from utility of the expected value, that is,

$$E[u(X)] \neq u(E[X]).$$

Similarly,

$$E\left[\frac{1}{X}\right] \neq \frac{1}{E[X]}.$$

In these cases the functions are linearized for further analysis. For example, let the discounted value of consumption in period $t + 1$ be βc_{t+1} where $0 < \beta < 1$ is the discount factor. Using the first-order Taylor approximation at c^*, the steady-state consumption level,

$$E_t\left[\frac{1}{\beta c_{t+1}}\right] \simeq E_t\left[\frac{1}{\beta}\left(\frac{1}{c^*} - \frac{1}{(c^*)^2}(c_{t+1} - c^*)\right)\right]$$

$$= \frac{1}{\beta c^*}\left(2 - \frac{E_t[c_{t+1}]}{c^*}\right).$$

Also, in general

$$E(X_1 X_2) \neq E(X_1)E(X_2)$$

unless X_1 and X_2 are statistically independent.

8.7 Stochastic Processes

A stochastic process is a sequence of random variables, $\{X_t\}$, $t = 0, 1, \ldots$. In this section we introduce several useful models in economics.

8.7.1 Martingales

A stochastic process is called a martingale if for $t = 0, 1, \ldots$,

1. $E[|X_t|] < \infty$,
2. $E[X_{t+1}|X_0, X_1, \ldots, X_t] = X_t$.

Taking expectation on both sides of condition 2 above gives

$$E[X_{t+1}] = E[X_t].$$

Therefore a martingale has constant mean. If all the X_t have independent and identical distributions, then the process is called a random walk.

8.7.2 Moving Average Models

A sequence of random variables $\{\epsilon_t\}$ is called a white noise process if for all $t \neq s$,

1. $E[\epsilon_t] = 0$,
2. $E[\epsilon_t^2] = \sigma^2$,
3. $\text{Cov}(\epsilon_t, \epsilon_s) = 0$.

A first-order moving average process, MA(1), is defined as

$$X_t = \mu + \epsilon_t + \theta\epsilon_{t-1},$$

where μ and θ are constants. The expected value of X_t in each period is

$$E[X_t] = E[\mu + \epsilon_t + \theta\epsilon_{t-1}] = \mu + E[\epsilon_t] + \theta E[\epsilon_{t-1}] = \mu.$$

The variance is

$$\begin{aligned}
E(X_t - \mu)^2 &= E(\epsilon_t + \theta\epsilon_{t-1})^2 \\
&= E(\epsilon_t^2 + 2\theta\epsilon_t\epsilon_{t-1} + \theta^2\epsilon_{t-1}^2) \\
&= \sigma^2 + 0 + \theta^2\sigma^2 \\
&= (1 + \theta^2)\sigma^2.
\end{aligned}$$

In general, a moving average process of order q, MA(q), is defined as

$$X_t = \mu + \epsilon_t + \theta_1\epsilon_{t-1} + \theta_2\epsilon_{t-2} + \cdots + \theta_q\epsilon_{t-q}. \tag{8.12}$$

8.7.3 Autoregressive Models

A first-order autoregressive process, AR(1), is defined as

$$X_t = c + \phi X_{t-1} + \epsilon_t.$$

This is in fact a first-order difference equation with non-constant coefficients as specified in Eq. (9.3), with $a = \phi$ and $b_{t-1} = c + \epsilon_t$. If $-1 < \phi < 1$, we can solve the equation backward using Eq. (9.5), that is,

$$\begin{aligned}
X_t &= \sum_{s=1}^{\infty} \phi^{s-1}(c + \epsilon_{t-s+1}) \\
&= \frac{c}{1-\phi} + \sum_{s=0}^{\infty} \phi^s \epsilon_{t-s}. \tag{8.13}
\end{aligned}$$

Finding the mean and the variance of this AR(1) process is left as an exercise. In this case the process has a constant mean and variance and is said to be **stationary**. On the other hand, if $\phi < -1$ or $\phi > 1$, the process is **non-stationary**. Comparing Eqs. (8.12) and (8.13) reveals that an AR(1) process can be expressed as a MA(∞) process with $\mu = c/(1 - \phi)$ and $\theta_i = \phi^i$. In general a stationary AR process of any order can be inverted into an MA process and vice versa.

It is often advantageous to use a mixed autoregressive and moving average model. For example, an ARMA(1, 1) process is defined as

$$X_t = c + \phi X_{t-1} + \epsilon_t + \theta \epsilon_{t-1}.$$

More on stochastic processes can be found in Taylor and Karlin (1998), Box, Jenkins, and Reinsel (1994) and Hamilton (1994). See Grinstead and Snell (1997), Hogg and Craig (1995), Rao (1973) or any book in mathematical probability and statistics for more detailed discussions on probability theory.

8.8   Exercises

1. In a random experiment, a coin is flipped until the first tail appears. Describe the sample space of the experiment.
2. In the game of craps, two dice are thrown and the sum of the number of dots on the dice is recorded. Describe the sample space of each throw.
3. Prove Theorem 8.1.
4. Show that the probability function defined in Eq. (8.1) satisfies the three axioms of probability measure.
5. Consider the experiment of drawing a card randomly from a deck of well-shuffled cards.
   (a) What is the probability of the outcome ♣10?
   (b) Find the probabilities of events A, B and $A \setminus B$ defined in Example 8.4.
6. Let $\Omega = \{a, b, c, d\}$. Determine if the following collections are σ-fields.
   (a) $\mathcal{F} = \{\varnothing, \Omega\}$
   (b) $\mathcal{F} = \{\varnothing, \{a\}, \{d\}, \{a, d\}, \{b, c\}, \Omega\}$
   (c) $\mathcal{F} = \{\varnothing, \{a, b\}, \{c, d\}, \Omega\}$
7. Show that the following collections are σ-fields of a sample space Ω:
   (a) $\{\varnothing, \Omega\}$.
   (b) $\{\varnothing, A, A^c, \Omega\}$, for any $A \subseteq \Omega$.
   (c) $\mathcal{P}(\Omega)$, the power set of Ω.
8. Show that if event A is independent of event B, then B is also independent of A.
9. (Wiggins 2006) A patient goes to see a doctor. The doctor performs a test with 99% reliability—that is, 99% of people who are sick test positive and 99% of the healthy people test negative. The doctor knows that only 1% of the people in the country are sick. If the patient tests positive, what are the chances the patient is sick?
10. (Paulos 2011) Assume that you're presented with three coins, two of them fair and the other a counterfeit that always lands heads. If you randomly pick one of the three coins, the probability that it's the counterfeit is 1 in 3. This is the prior probability of the hypothesis that the coin is counterfeit. Now after picking the coin, you flip it three times and observe that it lands heads each time. Seeing

this new evidence that your chosen coin has landed heads three times in a row, what is the revised posterior probability that it is the counterfeit?

11. Suppose that $\{B_1, B_2, \ldots, B_n\}$ is a partition of a sample space Ω. Prove the following **law of total probabilities**: For any event $A \subseteq \Omega$,

$$P(A) = \sum_{i=1}^{n} P(A|B_i)P(B_i).$$

12. Consider Example 8.5 of tossing a fair coin twice. Let X be the random variable of the number of times that tail shows up.
    (a) What is the support of X?
    (b) What is the probability mass function of X?
13. Prove Eq. (8.8).
14. Find the mean, variance and coefficient of variation of the random variable defined in Example 8.14.
15. Prove Theorem 8.2.
16. Prove Eq. (8.9).
17. The price of a new iTab computer is $500. It is known that a fraction d of the iTabs is defective. All consumers value the non-defective computers at $550 each. The same model of second-hand iTab can be found on eBay for $50, which includes shipping costs.
    (a) Do you expect to find non-defective iTabs for sale on eBay? Why or why not?
    (b) If the consumers are risk neutral, what is d?
18. A fair die is rolled 6 times. Let A be the event that the outcome is 5 each time. Let X be the number of times that event A occurs. Find the pmf, the mean and the variance of X.
19. Show that

$$\int_{-\infty}^{\infty} f_{2|1}(x_2|x_1) \, dx_2 = 1.$$

20. Let X_1 and X_2 have the joint pdf

$$f(x_1, x_2) = \begin{cases} 2, & 0 < x_1 < x_2 < 1, \\ 0 & \text{elsewhere.} \end{cases}$$

    (a) Find the marginal pdf of X_1 and X_2.
    (b) Find the conditional pdf of X_1 given $X_2 = x_2$.
    (c) Find the conditional mean of X_1 given $X_2 = x_2$.
21. Mr. Smith has two children, at least one of whom is a boy. What is the probability that the other is a boy? (Confused? See the discussion in Wikipedia.)

22. Consider the stock price model in Sect. 9.1.2. Suppose the dividends of the stock follow a random walk, that is,

$$d_t = d_{t-1} + e_t,$$

where e_t has an independent and identical distribution in every period with $E[e_t] = 0$. Instead of the adaptive expectation as in Eq. (9.7), the investors have rational expectations, $E_t[p_{t+1}] = p_{t+1}$.

(a) Derive the stock price p_t.

(b) Repeat the analysis if the dividends are constant, that is, $d_t = d$ for all t.

23. Find the mean and variance of the AR(1) process if $-1 < \phi < 1$.

24. Consider the following passage from Grinstead and Snell (1997, p. 405):

> Most of our study of probability have dealt with independent trials processes. These processes are the basis of classical probability theory and much of statistics. We have discussed two of the principal theorems for these processes: the Law of Large Numbers and the Central Limit Theorem.
>
> We have seen that when a sequence of chance experiments forms an independent trials process, the possible outcomes for each experiment are the same and occur with the same probability. Further, knowledge of the outcomes of the previous experiments does not influence our predictions for the outcomes of the next experiment. The distribution for the outcomes of a single experiment is sufficient to construct a tree and a tree measure for a sequence of n experiments, and we can answer any probability question about these experiments by using this tree measure.

Rewrite the second paragraph in the mathematical language of probability theory.

References

Artin, E. (1964). *The gamma function*. New York: Holt, Rinehart and Winston.

Box, G. E. P., Jenkins, G. M., & Reinsel, G. C. (1994). *Time series analysis: forecasting and control*, Third edition. Englewood Cliffs: Prentice-Hall.

Grinstead, C. M., & Snell, J. L. (1997). *Introduction to probability*, Second Revised Edition. Providence: American Mathematical Society. (Downloadable from the Chance web site).

Economist. (1999). Getting the Goat, February 18 issue.

Hamilton, J. D. (1994). *Time series analysis*. Princeton: Princeton University Press.

Hogg, R. V., & Craig, A. T. (1995). *Introduction to mathematical statistics*, Fifth Edition. Englewood: Prentice-Hall, Inc.

Paulos, J. A. (2011). The Mathematics of Changing Your Mind. *The New York Times*, August 5 issue.

Rao, C. R. (1973). *Linear statistical inference and its applications*, Second Edition. New York: John Wiley & Sons.

Ross, J. F. (2004). Pascal's legacy. *EMBO Reports*, Vol. 5, Special issue, S7–S10.

Taylor, H. M., & Karlin, S. (1998). *An introduction to stochastic modeling*, Third Edition. San Diego: Academic Press.

Timoshenko, S. (1940). *Strength of materials*. New York: D. Van Nostrand Company.

Tversky, A., & Kahneman, D. (1974). Judgment under uncertainty: heuristics and biases. *Science, 185*(4157), 1124–1131.

Wiggins, C. (2006). Bayes's theorem. *Scientific American*, December 4 issue.

Dynamic Modelling

9

In this chapter we extend economic modelling to include the time dimension. Dynamic modelling is the essence of macroeconomic theory. Our discussions provide the basic ingredients of the so-called dynamic general equilibrium model.

9.1 First-Order Difference Equation

9.1.1 Constant Coefficients

Suppose that $\{x_t\}$ is a sequence of variables in $\mathbb{R}$ which satisfies the first-order difference equation

$$x_{t+1} = ax_t + b, \quad t = 0, 1, 2, \ldots, \tag{9.1}$$

where a and b are constants and $a \neq 0$. If $b = 0$ the equation is called homogeneous. If the value of x_t is known in period t, then all subsequent terms can be expressed in terms of x_t:

$$x_{t+n} = \begin{cases} a^n x_t + b\dfrac{1 - a^n}{1 - a} & \text{if } a \neq 1 \\ x_t + bn & \text{if } a = 1 \end{cases} \tag{9.2}$$

for $n = 1, 2, \ldots$. This can be proven by mathematical induction. We first show that Eq. (9.2) is true for $n = 1$. Then by assuming that (9.2) is true for a particular period n, we show that it is also true in period $n + 1$.

For the case of $a \neq 1$, it is obvious that putting $n = 1$ in Eq. (9.2) reduces it to Eq. (9.1). Now assume that Eq. (9.2) holds for x_{t+n}, then

© Springer Nature Switzerland AG 2019
K. Yu, *Mathematical Economics*, Springer Texts in Business and Economics,
https://doi.org/10.1007/978-3-030-27289-0_9

$$ax_{t+n} + b = a\left(a^n x_t + b\frac{1-a^n}{1-a}\right) + b$$

$$= a^{n+1}x_t + ab\frac{1-a^n}{1-a} + b$$

$$= a^{n+1}x_t + b\left(a\frac{1-a^n}{1-a} + 1\right)$$

$$= a^{n+1}x_t + b\left(\frac{a(1-a^n) + (1-a)}{1-a}\right)$$

$$= a^{n+1}x_t + b\left(\frac{a - a^{n+1} + 1 - a}{1-a}\right)$$

$$= a^{n+1}x_t + b\left(\frac{1 - a^{n+1}}{1-a}\right)$$

$$= x_{t+n+1},$$

which shows that Eq. (9.1) holds for x_{t+n+1}. The proof for the case $a = 1$ is left as an exercise.

If $-1 < a < 1$, then $\lim_{n\to\infty} a^n = 0$. By Eq. (9.2) we have

$$\lim_{n\to\infty} x_{t+n} = \frac{b}{1-a}.$$

That is, if $-1 < a < 1$, x_t converges to a steady-state value of $b/(1-a)$. Figure 9.1 plots the values of x_t with $a = 0.3, b = 10$, with initial value $x_0 = 1$. The series quickly converges to the steady-state value of $10/(1-0.3) = 14.286$ within ten periods.

Fig. 9.1 First-order difference equation

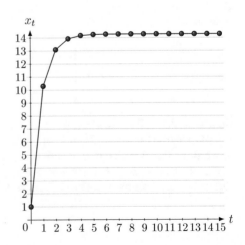

On the other hand, if $a \geq 1$ or $a \leq -1$, then x_t diverges and does not achieve a steady state.

Alternatively, Eq. (9.1) can be seen as a dynamical system, with

$$x_{t+1} = f(x_t) = ax_t + b.$$

A steady-state value for x_t is a fixed point of the function f. Thus by setting $x_{t+1} = x_t = x^*$, we have

$$x^* = \frac{b}{1 - a}$$

as above. The system will converge to the fixed point x^* if $|f'(x^*)| < 1$ and will diverge if $|f'(x^*)| > 1$. See Devaney (2003, p. 24) for details.

Figure 9.2a illustrates the case when $0 < a < 1$. The initial value of x is equal to x_0. The value of $x_1 = ax_0 + b$ is reflected back to the x-axis by the $45°$ line and the process goes on. Eventually x converges to the fixed point at x^*. Starting points with values greater than x^* go through a similar process converging to x^*. In Fig. 9.2b the coefficient a is greater than one. Values of x not equal to the fixed point x^* will move away from it. You should draw the cases of $-1 < a < 0$ and $a < -1$ to reveal the dynamics of the so-called cobweb models.

9.1.2 Non-constant Coefficients

Suppose that b is not a constant but a sequence. Then (9.1) becomes

$$x_{t+1} = ax_t + b_t, \quad t = 0, 1, 2, \ldots, \tag{9.3}$$

If $a \geq 1$ or $a \leq -1$, this difference equation can be solved "forward" to give the value of x_t, that is,

Fig. 9.2 Convergence of first-order difference equation. (**a**) $0 < a < 1$. (**b**) $a > 1$

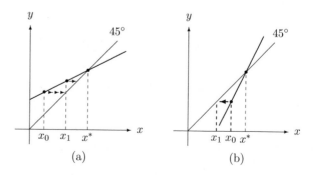

$$x_t = \frac{1}{a}(-b_t + x_{t+1})$$

$$= \frac{1}{a}\left(-b_t + \frac{1}{a}(-b_{t+1} + x_{t+2})\right)$$

$$= \frac{1}{a}\left(-b_t + \frac{1}{a}\left[-b_{t+1} + \frac{1}{a}(-b_{t+2} + x_{t+3})\right]\right)$$

$$\vdots$$

$$= -\frac{1}{a}\left(b_t + \frac{b_{t+1}}{a} + \frac{b_{t+2}}{a^2} + \cdots\right) + \lim_{s\to\infty}\frac{x_{t+s}}{a^s}$$

$$= -\frac{1}{a}\sum_{s=0}^{\infty}\frac{b_{t+s}}{a^s} + \lim_{s\to\infty}\frac{x_{t+s}}{a^s}$$

$$= -\sum_{s=0}^{\infty}\frac{b_{t+s}}{a^{s+1}} + \lim_{s\to\infty}\frac{x_{t+s}}{a^s}.$$

Convergence of this series depends on the behaviours of the sequence $\{b_t\}$. In particular, a necessary condition for convergence is the so-called transversality condition,

$$\lim_{s\to\infty}\frac{x_{t+s}}{a^s} = 0.$$

If the transversality condition is satisfied, then

$$x_t = -\sum_{s=0}^{\infty}\frac{b_{t+s}}{a^{s+1}}. \tag{9.4}$$

For the case of $-1 < a < 1$, we can solve the difference equation (9.3) "backward" to get

$$x_t = \sum_{s=1}^{\infty}a^{s-1}b_{t-s}. \tag{9.5}$$

In the special case that $b_{t+s} = b$ for all s, Eq. (9.3) becomes a geometric series and so

$$x_t = \frac{b}{1-a}.$$

See Goldberg (1986) for more on difference equations. Chapter 2 in Hamilton (1994) also contains discussions on difference equations using lag operators.

Example 9.1 (Azariadis 1993, Chapter 3) Suppose that investors have two choices of asset. The first is a riskless government bond that pays an interest at the rate of r per period. The second is a stock that pays dividend d_t in period t. The stock is bought or sold at market price p_t before the dividend is paid out. Then a risk-neutral investor faces

$$(1+r)p_t = d_t + E_t[p_{t+1}], \tag{9.6}$$

that is, an amount of p_t invested in government bond should be equal to the expected return invested in the stock. The investor has adaptive expectation in every period:

$$E_t[p_{t+1}] = \lambda p_t + (1-\lambda)E_{t-1}[p_t], \tag{9.7}$$

with $0 < \lambda < 1$. This means that forecasted price $E_t[p_{t+1}]$ is partially adjusted by a factor of λ towards the forecasting error $p_t - E_{t-1}[p_t]$ in the last period. Substitute the expectation in Eq. (9.6) into (9.7), we get

$$(1+r)p_t - d_t = \lambda p_t + (1-\lambda)[(1+r)p_{t-1} - d_{t-1}].$$

Rearranging and shifting one period forward to get

$$p_{t+1} = ap_t + b_t,$$

where

$$a = \frac{(1+r)(1-\lambda)}{1+r-\lambda},$$

and

$$b_t = \frac{d_{t+1} - (1-\lambda)d_t}{1+r-\lambda}.$$

Given that $r > 0$, it is clear that $0 < a < 1$. We can therefore apply Eq. (9.5) to obtain the price of the stock in period t,

$$p_t = \sum_{s=1}^{\infty} a^{s-1} b_{t-s}.$$

In the special case that dividends are constant, that is, $d_t = d$ for all t,

$$b_t = b = \frac{\lambda d}{1+r-\lambda}$$

for all t. The price of the stock becomes

$$p_t = \sum_{s=1}^{\infty} a^{s-1} b = \frac{b}{1-a} = \frac{1+r-\lambda}{\lambda r} \frac{\lambda d}{1+r-\lambda} = \frac{d}{r}.$$

The price/earning ratio of the U.S. stock market, which is p_t/d_t in our model, has fluctuated over the years. In recent years the average value is about 20.

In passing, we should mention that higher-order difference equations can be transformed into first-order vector difference equations. For example, consider a linearly homogeneous second-order equation

$$x_{t+1} = ax_t + bx_{t-1}. \tag{9.8}$$

We can let $y_t = x_{t-1}$ so that the system becomes

$$\begin{bmatrix} x_{t+1} \\ y_{t+1} \end{bmatrix} = \begin{bmatrix} a & b \\ 1 & 0 \end{bmatrix} \begin{bmatrix} x_t \\ y_t \end{bmatrix}.$$

We shall discuss linear dynamical system in Sect. 9.4.1.

9.2   Dynamical Systems

A general dynamical system can be specified as

$$\mathbf{x}_{t+1} = f_t(\mathbf{x}_t, \mathbf{u}_t), \tag{9.9}$$

$$\mathbf{y}_t = g_t(\mathbf{x}_t, \mathbf{u}_t), \tag{9.10}$$

where

- $t \in \mathbb{Z} = \{\dots, -2, -1, 0, 1, 2, \dots\}$ denotes discrete time periods,
- $\mathbf{x}_t \in \mathbb{R}^n$ is the state variable,
- $\mathbf{u}_t \in \mathbb{R}^m$ is the input or control variable,
- $\mathbf{y}_t \in \mathbb{R}^p$ is the output variable,
- $f_t : \mathbb{R}^n \times \mathbb{R}^m \to \mathbb{R}^n$ is the transition function,
- $g_t : \mathbb{R}^n \times \mathbb{R}^m \to \mathbb{R}^p$ is the payoff function, sometimes called the criterion or return function.

The transition equation (9.9) relates the state variable in period $t+1$ to the state and control variables in period t. The function f_t is usually governed by some physical laws or technology. In an optimal control problem, the controller chooses the sequence $\{\mathbf{u}_t\}$ to get a desirable sequence of outputs $\{\mathbf{y}_t\}$ in Eq. (9.10). When there is no control variable, the system is called **autonomous**. In many applications, f_t and g_t are time-invariant or **stationary** and are therefore written without the time subscripts.

Example 9.2 Consider a simple Ramsey model in economics. The state variable is the capital stock k_t and the control variable is consumption c_t. The transition equation is

$$k_{t+1} = F(k_t) + (1 - \delta)k_t - c_t, \tag{9.11}$$

where F is the aggregate production function and δ is the depreciation rate of capital. The output is the utility of the aggregate household so that the payoff function in each period is $U(c_t)$. The objective of the central planner is to maximize the discounted values of present and future utility, that is,

$$\max_{c_t} \sum_{t=0}^{\infty} \beta^t U(c_t),$$

where $\beta = 1/(1 + \theta)$ is the social discount factor. The following Euler equation is obtained from the necessary conditions of optimization (see Sect. 9.3):

$$\beta \frac{U'(c_{t+1})}{U'(c_t)} \left[F'(k_{t+1}) + 1 - \delta \right] = 1. \tag{9.12}$$

Assuming the utility function U is strictly concave so that $U''(c) < 0$ for all values of c, the Euler equation can be solved for c_{t+1} as a function of c_t and k_t (the term k_{t+1} can be substituted by c_t and k_t using Eq. (9.11)). The result is a two-dimensional autonomous dynamical system with $\mathbf{x}_t = [k_t \ c_t]^T$.

Consider an autonomous and stationary dynamical system with given initial state $\mathbf{x}$. After two periods we have $f(f(\mathbf{x})) = f^2(\mathbf{x})$. In general, $f^{t+s}(\mathbf{x}) = f^s(f^t(\mathbf{x}))$. The sequence $\{\mathbf{x}_t\}$ for $t \in \mathbb{Z}$ is called the **orbit** of $\mathbf{x}$.[1] When the sequence only includes $t = \{0, 1, 2, \ldots\}$, it is called a **forward orbit**. Similarly, a **backward orbit** consists of $t = \{0, -1, -2, \ldots\}$. A **fixed point p** is defined by $f(\mathbf{p}) = \mathbf{p}$. The set of all fixed points of f is written as Fix(f). In economics a fixed point is generally called an equilibrium.

In dynamic general equilibrium models, the system is sometimes linearized by first-order Taylor approximation about the steady-state values. An autonomous linear dynamical system can be written as

$$\mathbf{x}_{t+1} = A\mathbf{x}_t + B\mathbf{u}_t,$$

$$\mathbf{y}_t = C\mathbf{x}_t + D\mathbf{u}_t,$$

where A, B, C and D are matrices with the conforming dimensions.[2]

---

[1] Orbits are often called trajectories or flows.

[2] Boyd (2008) provides a detailed analysis of linear dynamical systems.

9.3   Dynamic Programming

In this section we assume that the payoff function g is differentiable and real-valued. That is, the output variable $y_t \in \mathbb{R}$.

9.3.1   Basic Theory

A typical dynamic programming problem is

$$\max_{\mathbf{u}_t} \sum_{t=0}^{\infty} \beta^t g(\mathbf{x}_t, \mathbf{u}_t), \tag{9.13}$$

subject to the transition equation

$$\mathbf{x}_{t+1} = f(\mathbf{x}_t, \mathbf{u}_t), \tag{9.14}$$

where $\beta \in (0, 1]$ is a discount factor. The value of the state variable in period 0, $\mathbf{x}_0$, is given. In other words, we want to maximize the sum of the present values of the output variables from period $t = 0$ onward. Notice that the payoff function g and the transition function f are stationary. This will allow us to express the system in a recursive form below. Often we assume that g is differentiable and concave and the feasible set $S = \{(\mathbf{x}_{t+1}, \mathbf{x}_t) : \mathbf{x}_{t+1} \leq f(\mathbf{x}_t, \mathbf{u}_t), t = 0, 1, \ldots\}$ is convex and compact.

If the solution of the above problem exists, $\mathbf{u}_t = h(\mathbf{x}_t)$ is called the **policy function**. With the initiate state $\mathbf{x}_0$ and the transition equation (9.14), the sequence $\{(\mathbf{x}_t, \mathbf{u}_t)\}$ is obtained. By putting the sequence $\{(\mathbf{x}_t, \mathbf{u}_t)\}$ in the objective function in (9.13), we get the **value function**

$$v(\mathbf{x}_0) = \max_{\mathbf{u}_t} \left\{ \sum_{t=0}^{\infty} \beta^t g(\mathbf{x}_t, \mathbf{u}_t) : \mathbf{x}_{t+1} = f(\mathbf{x}_t, \mathbf{u}_t) \right\}.$$

After some time periods, say s, we obtain the state variable $\mathbf{x}_s$. The structure of the problem remains the same since g and f are stationary. Therefore the value function, now depends on $\mathbf{x}_s$, is

$$v(\mathbf{x}_s) = \max_{\mathbf{u}_t} \left\{ \sum_{t=s}^{\infty} \beta^{t-s} g(\mathbf{x}_t, \mathbf{u}_t) : \mathbf{x}_{t+1} = f(\mathbf{x}_t, \mathbf{u}_t) \right\}.$$

In general, in any period t, the value function is the sum of current payoff function and the next period's value function, that is,

$$v(\mathbf{x}_t) = \max_{\mathbf{u}_t} \{g(\mathbf{x}_t, \mathbf{u}_t) + \beta v(\mathbf{x}_{t+1}) : \mathbf{x}_{t+1} = f(\mathbf{x}_t, \mathbf{u}_t)\}$$

$$= \max_{\mathbf{u}_t} \{g(\mathbf{x}_t, \mathbf{u}_t) + \beta v(f(\mathbf{x}_t, \mathbf{u}_t))\}. \tag{9.15}$$

In fact, since the value function has the same structure in every period, we can express the above equation recursively as

$$v(\mathbf{x}) = \max_{\mathbf{u}} \{g(\mathbf{x}, \mathbf{u}) + \beta v(f(\mathbf{x}, \mathbf{u}))\}. \tag{9.16}$$

We can substitute the maximizer $\mathbf{u}$ with the policy function $h(\mathbf{x})$ in the above equation to obtain

$$v(\mathbf{x}) = g(\mathbf{x}, h(\mathbf{x})) + \beta v(f(\mathbf{x}, h(\mathbf{x}))). \tag{9.17}$$

Equations (9.15), (9.16) and (9.17) are different versions of the **Bellman equation**. Given g and f, Eq. (9.17) is a functional equation with unknown functions v and h to be found.

Existence of the solution is confirmed by the following theorem. The proof can be found in Stokey and Lucas (1989, p. 79) and Ljungqvist and Sargent (2004, p. 1011–1012).

Theorem 9.1 *Suppose that g is continuous, concave and bounded, $\beta \in (0, 1)$, and the feasible set S is nonempty, compact and convex. Then there exists a unique function v that solves the Bellman equation (9.17).*

Methods for solving the Bellman equation for the policy function h and value function v include iterations of h, iterations of v, the method of undetermined coefficients and numerical analysis using computers. You can consult Stokey and Lucas (1989) or Ljungqvist and Sargent (2004) for details. In what follows we assume that the value function v is differentiable and derive the first-order conditions for maximizing the Bellman equation. Then we apply the envelope theorem to the value function v. Together we obtain the same set of necessary conditions as in the Lagrange method.

Let $\lambda_t = \nabla v(\mathbf{x}_t)$ and $\lambda_{t+1} = \nabla v(\mathbf{x}_{t+1})$. The first-order condition of the maximization problem in (9.15) is

$$\nabla_{\mathbf{u}} g(\mathbf{x}_t, \mathbf{u}_t) + \beta D_{\mathbf{u}} f(\mathbf{x}_t, \mathbf{u}_t)^{\mathrm{T}} \lambda_{t+1} = \mathbf{0} \in \mathbb{R}^m. \tag{9.18}$$

Next, we differentiate the functional equation (9.17) with respect to the state variable $\mathbf{x}$:

$$\lambda_t = \nabla v(\mathbf{x}_t) = \nabla_{\mathbf{x}} g + (Dh)^{\mathrm{T}} \nabla_{\mathbf{u}} g + \beta[(D_{\mathbf{x}} f)^{\mathrm{T}} + (Dh)^{\mathrm{T}} (D_{\mathbf{u}} f)^{\mathrm{T}}] \lambda_{t+1}$$

$$= \nabla_{\mathbf{x}} g + \beta(D_{\mathbf{x}} f)^{\mathrm{T}} \lambda_{t+1} + (Dh)^{\mathrm{T}} [\nabla_{\mathbf{u}} g + \beta(D_{\mathbf{u}} f)^{\mathrm{T}} \lambda_{t+1}].$$

Table 9.1 Dimensions of
derivatives

Derivatives	Dimensions
λ_t, λ_{t+1}	$n \times 1$
$\nabla_{\mathbf{u}} g(\mathbf{x}_t, \mathbf{u}_t)$	$m \times 1$
$\nabla_{\mathbf{x}} g(\mathbf{x}_t, \mathbf{u}_t)$	$n \times 1$
$D_{\mathbf{u}} f(\mathbf{x}_t, \mathbf{u}_t)$	$n \times m$
$D_{\mathbf{x}} f(\mathbf{x}_t, \mathbf{u}_t)$	$n \times n$
$Dh(\mathbf{x}_t)$	$m \times n$

By the first-order condition (9.18), the term inside the square bracket in the last equality above is the zero vector in $\mathbb{R}^m$. Therefore we have

$$\lambda_t = \nabla_{\mathbf{x}} g(\mathbf{x}_t, \mathbf{u}_t) + \beta D_{\mathbf{x}} f(\mathbf{x}_t, \mathbf{u}_t)^{\mathrm{T}} \lambda_{t+1} \in \mathbb{R}^n. \tag{9.19}$$

Table 9.1 keeps track of the dimensions of the derivatives in the above algebra. You should verify that all matrix multiplications in applying the chain rule are compatible.

Equation (9.19) is similar to the envelope theorem in static optimization. It effectively means that the rate of change (gradient) of the value function v with respect to the state variable $\mathbf{x}_t$ is equal to the partial derivative of the Bellman equation with respect to $\mathbf{x}_t$.

9.3.2 Recipe

Here is a set of cookbook style procedures to follow as an alternative method to the Lagrange method in analysing dynamic optimization problems:

1. Make sure that the objective function is additively separable for each period and the payoff function $g(\mathbf{x}_t, \mathbf{u}_t)$ is stationary.
2. Make sure that the transition function is stationary and expressed in the form

$$\mathbf{x}_{t+1} = f(\mathbf{x}_t, \mathbf{u}_t). \tag{9.20}$$

3. Set up the Bellman equation

$$v(\mathbf{x}_t) = \max_{\mathbf{u}_t} \{ g(\mathbf{x}_t, \mathbf{u}_t) + \beta v(f(\mathbf{x}_t, \mathbf{u}_t)) \}.$$

4. Let $\lambda_t = \nabla v(\mathbf{x}_t)$ and $\lambda_{t+1} = \nabla v(\mathbf{x}_{t+1})$.
5. Differentiate the right-hand side of the Bellman equation with respect to the control variable $\mathbf{u}_t$ and set the gradient equal to the zero vector $\mathbf{0} \in \mathbb{R}^m$ as in Eq. (9.18).
6. Differentiate the right-hand side of the Bellman equation with respect to the state variable $\mathbf{x}_t$ and set the gradient equal to $\lambda_t \in \mathbb{R}^n$ as in Eq. (9.19).

7. The results in Steps 5 and 6 together with the transition equation (9.20) form the set of necessary conditions for optimization.

Example 9.3 Consider the Ramsey model in Example 9.2. In this case $m = n = 1$. The Bellman equation is

$$v(k_t) = \max_{c_t}\{U(c_t) + \beta v[F(k_t) + (1 - \delta)k_t - c_t]\}.$$

The necessary conditions from Steps 5 and 6 above are

$$U'(c_t) - \beta\lambda_{t+1} = 0,$$
$$\beta[F'(k_{t+1}) + 1 - \delta]\lambda_{t+1} = \lambda_t.$$

From the first equation $\lambda_{t+1} = U'(c_t)/\beta$ so that $\lambda_t = U'(c_{t-1})/\beta$. Substituting these into the second equation above and shifting one period forward give the Euler equation in (9.12).

The Ramsey model above can also be expressed in continuous time as

$$\max_{c(t)} \int_0^\infty e^{-\theta t} U(c(t))\, dt,$$

subject to

$$\frac{dk(t)}{dt} = F(k(t)) - \delta k(t) - c(t).$$

The problem can be solved using the technique of optimal control theory and is popular among theorists of economic growth and natural resource economics. The results given by the continuous time version are similar to the discrete time version. But since economic data such as output, investment, consumption and inflation come in discrete form, the dynamic programming approach is more convenient in empirical analysis. An excellent exposition of optimal control theory and its applications in economics can be found in Weitzman (2003).

Example 9.4 In this example we analyse the dynamics of durable goods set up in Wickens (2011, p. 74–76) using dynamic programming. Utility $U(c_t, D_t)$ in each period depends on consumption of nondurable goods c_t and the stock of durable goods D_t. The two state variables are D_t and asset holding a_t. The two control variables are c_t and the purchase of durable goods in each period, d_t, which has a relative price equal to p_t. The transition equations are the durable goods accumulation and the budget constraint:

$$D_{t+1} = d_t + (1 - \delta)D_t,$$
$$a_{t+1} = x_t + (1 + r_t)a_t - c_t - p_t d_t,$$

where δ is the depreciation rate, x_t is exogenous income and r_t is the interest rate. The Bellman equation is

$$v(D_t, a_t) = \max_{c_t, d_t} \{U(c_t, D_t) + \beta v[d_t + (1 - \delta)D_t, x_t + (1 + r_t)a_t - c_t - p_t d_t]\}.$$

Let $\nabla v(D_t, a_t) = (\lambda_{D,t} \ \lambda_{a,t})^{\mathrm{T}}$. Equation (9.18) is

$$\begin{bmatrix} U_{c,t} \\ 0 \end{bmatrix} + \beta \begin{bmatrix} 0 & -1 \\ 1 & -p_t \end{bmatrix} \begin{bmatrix} \lambda_{D,t+1} \\ \lambda_{a,t+1} \end{bmatrix} = \begin{bmatrix} 0 \\ 0 \end{bmatrix}, \tag{9.21}$$

and Eq. (9.19) is

$$\begin{bmatrix} U_{D,t} \\ 0 \end{bmatrix} + \beta \begin{bmatrix} 1 - \delta & 0 \\ 0 & 1 + r_t \end{bmatrix} \begin{bmatrix} \lambda_{D,t+1} \\ \lambda_{a,t+1} \end{bmatrix} = \begin{bmatrix} \lambda_{D,t} \\ \lambda_{a,t} \end{bmatrix}, \tag{9.22}$$

The square matrix in Eq. (9.21) can be inverted to get

$$\begin{bmatrix} \lambda_{D,t+1} \\ \lambda_{a,t+1} \end{bmatrix} = \frac{1}{\beta} \begin{bmatrix} p_t U_{c,t} \\ U_{c,t} \end{bmatrix}.$$

Substituting this into Eq. (9.22), we have

$$\begin{bmatrix} U_{D,t} \\ 0 \end{bmatrix} + \begin{bmatrix} 1 - \delta & 0 \\ 0 & 1 + r_t \end{bmatrix} \begin{bmatrix} p_t U_{c,t} \\ U_{c,t} \end{bmatrix} = \frac{1}{\beta} \begin{bmatrix} p_{t-1} U_{c,t-1} \\ U_{c,t-1} \end{bmatrix}.$$

The second line in the above equation gives the Euler equation. The first line, using the Euler equation, can be expressed as

$$U_{D,t} = \frac{U_{c,t-1}}{\beta} \left(p_{t-1} - \frac{1 - \delta}{1 + r_t} p_t \right).$$

See Adda and Cooper (2003, Chapter 7) for a survey on the economics of durable consumption.

9.3.3 Linear Quadratic Models

A popular class of dynamic programming models is the linear quadratic model. They are useful in solving rational expectation models and as second-order approximations to nonlinear models. The structure of the model is that the payoff function g is the sum of two quadratic forms of the state variable x_t and the control variables u_t, while the transition function f is a linear function of x_t and u_t. The problem is

$$\max_{\mathbf{u}_t} \sum_{t=0}^{\infty} -\beta^t \left(\mathbf{x}_t^{\mathrm{T}} R \mathbf{x}_t + \mathbf{u}_t^{\mathrm{T}} Q \mathbf{u}_t \right), \quad \beta \in (0, 1),$$

subject to

$$\mathbf{x}_{t+1} = A\mathbf{x}_t + B\mathbf{u}_t, \quad \mathbf{x}_0 \text{ given.}$$

Here R and Q are positive semidefinite symmetric matrices of dimensions $n \times n$ and $m \times m$ respectively, A is an $n \times n$ matrix and B is an $n \times m$ matrix. The Bellman equation is

$$v(\mathbf{x}_t) = \max_{\mathbf{u}_t} \left\{ - \left(\mathbf{x}_t^{\mathrm{T}} R \mathbf{x}_t + \mathbf{u}_t^{\mathrm{T}} Q \mathbf{u}_t \right) + \beta v(A\mathbf{x}_t + B\mathbf{u}_t) \right\}.$$

Writing $\lambda_t = \nabla v(\mathbf{x}_t)$ and $\lambda_{t+1} = \nabla v(\mathbf{x}_{t+1})$, the necessary conditions are

$$- 2Q\mathbf{u}_t + \beta B^{\mathrm{T}}\lambda_{t+1} = \mathbf{0} \in \mathbb{R}^m, \tag{9.23}$$

$$-2R\mathbf{x}_t + \beta A^{\mathrm{T}}\lambda_{t+1} = \lambda_t \in \mathbb{R}^n. \tag{9.24}$$

Premultiplying Eq. (9.23) by B, we get

$$\beta B B^{\mathrm{T}}\lambda_{t+1} = 2B Q\mathbf{u}_t.$$

Assume that $m \geq n$ and B has rank n. Then BB^{T} is an $n \times n$ invertible matrix so that

$$\lambda_{t+1} = 2 \left[\beta B B^{\mathrm{T}} \right]^{-1} B Q\mathbf{u}_t,$$

and

$$\lambda_t = 2 \left[\beta B B^{\mathrm{T}} \right]^{-1} B Q\mathbf{u}_{t-1}.$$

Substituting λ_t and λ_{t+1} into Eq. (9.24) and rearranging, we obtain

$$A^{\mathrm{T}} \left[\beta B B^{\mathrm{T}} \right]^{-1} B Q\mathbf{u}_t - \left[\beta B B^{\mathrm{T}} \right]^{-1} B Q\mathbf{u}_{t-1} = R\mathbf{x}_t \in \mathbb{R}^n.$$

This is a form of Euler equation which gives the intertemporal relation of the control variable $\mathbf{u}$. If $m > n$, we need more constraints among the components of $\mathbf{u}$ such as government budget constraint and money supply rule.

9.3.4   Finite Horizon Problems

In some models the optimization problem has finite time periods. Such cases can arise in business contracts, retirement or natural resource extraction. Suppose that the terminating time period is T, the dynamic programming problem becomes

$$\max_{\mathbf{u}_t} \sum_{t=0}^{T-1} \beta^t g(\mathbf{x}_t, \mathbf{u}_t) + \beta^T v(\mathbf{x}_T),$$

subject to

$$\mathbf{x}_{t+1} = f(\mathbf{x}_t, \mathbf{u}_t), \quad t = 0, \ldots, T - 1.$$

As before the initial value of the state variable, $\mathbf{x}_0$, is given. The additional term in the objective function, $\beta^T v(\mathbf{x}_T)$, is the discounted scrap value, which is usually specified in the model. In fact, the problem can be solved by starting from the last period and working backward. The following example is taken from Carter (2001, Chapter 7).[3] It illustrates an application of dynamic programming in operations management. The structure is similar to the "cake eating problem" described in Adda and Cooper (2003, p. 14–15).

Example 9.5 A mining company has acquired a licence to extract a rare earth element for three years. The geophysicist of the company has estimated that there are 128 tonnes of the ore in the mine. The market price of the ore is $1 million per tonne. The total cost of extraction is u_t^2/x_t, where u_t is the rate of extraction and x_t is the stock of the ore remaining in the ground. The discount rate β is assumed to be 1. The problem can be set up as

$$\max_{u_t} \sum_{t=0}^{3} \left(u_t - \frac{u_t^2}{x_t} + v(x_3) \right),$$

subject to

$$x_{t+1} = x_t - u_t, \quad t = 0, 1, 2.$$

Since any quantity of ore remains in the ground after year 3 has no value to the company, we can assume that $v(x_3) = 0$. The Bellman equation in period $t = 2$ is

$$v(x_2) = \max_{u} \left\{ u - \frac{u^2}{x_2} \right\},$$

---

[3] See also Conrad (2010, p. 55).

which is maximized when $u = x_2/2$. Hence

$$v(x_2) = \frac{x_2}{2} - \frac{x_2^2}{4x_2} = \frac{x_2}{4}.$$

The Bellman equation in period $t = 1$ is

$$v(x_1) = \max_u \left\{ u - \frac{u^2}{x_1} + v(x_2) \right\}$$

$$= \max_u \left\{ u - \frac{u^2}{x_1} + \frac{1}{4}(x_1 - u) \right\}.$$

The first-order condition is

$$1 - \frac{2u}{x_1} - \frac{1}{4} = 0,$$

which gives $u = 3x_1/8$ so that

$$v(x_1) = \frac{3x_1}{8} - \frac{9x_1^2}{64x_1} + \frac{1}{4}\left(x_1 - \frac{3x_1}{8} \right) = \frac{25}{64}x_1.$$

Finally, the Bellman equation for $t = 0$ is

$$v(x_0) = \max_u \left\{ u - \frac{u^2}{x_0} + v(x_1) \right\}$$

$$= \max_u \left\{ u - \frac{u^2}{x_0} + \frac{25}{64}(x_0 - u) \right\}.$$

The first-order condition is

$$1 - \frac{2u}{x_0} - \frac{25}{64} = 0,$$

which gives $u = 39x_0/128$. The following table summarizes the stock and the optimal extraction rate in each period.

t	x_t	u_t
0	128	39
1	89	$\frac{3}{8} \times 89 = 33.4$
2	$\frac{5}{8} \times 89 = 55.6$	$\frac{5}{16} \times 89 = 27.8$

Notice that the cost of extraction is inversely related to the stock, so it is not possible to extract all the mineral in the ground.

9.4   Autonomous Dynamical Systems

In this section we consider autonomous systems with no control variable or when the control variables follow some necessary conditions of an optimization process.

9.4.1   Linear Systems

Consider an autonomous linear dynamic system with no control variable. Assume that the linear transition function f is symmetric. Then f can be represented by a symmetric $n \times n$ square matrix A relative to the standard basis in $\mathbb{R}^n$, that is, the system is simply

$$\mathbf{x}_{t+1} = A\mathbf{x}_t.$$

By the spectral theorem, A can be written as

$$A = P \Lambda P^{\mathrm{T}}$$

where $\Lambda = \mathrm{diag}(\lambda_1, \lambda_2, \ldots, \lambda_n)$, that is, Λ is a $n \times n$ diagonal matrix with the eigenvalues at the principal diagonal and the off-diagonal elements are all zero. The square matrix $P = (\mathbf{v}_1 \ \mathbf{v}_2 \ \cdots \ \mathbf{v}_n)$ is an orthonormal matrix with columns formed by the corresponding normalized eigenvectors of A. The transition equation becomes

$$\mathbf{x}_{t+1} = P \Lambda P^{\mathrm{T}} \mathbf{x}_t.$$

Geometrically, the vector $\mathbf{x}_t$ is projected onto a different coordinate system by P^{T} with $\{\mathbf{v}_1, \mathbf{v}_2, \ldots, \mathbf{v}_n\}$ as the basis. Then each projected coordinate of $\mathbf{x}_t$ is multiplied by the corresponding eigenvalue λ_i. The rescaled coordinates are then transformed back by P to the coordinates relative to the standard basis in $\mathbb{R}^n$.

In period $t + 2$, the state variable becomes

$$\begin{aligned}
\mathbf{x}_{t+2} &= A\mathbf{x}_{t+1} \\
&= A^2 \mathbf{x}_t \\
&= (P \Lambda P^{\mathrm{T}})(P \Lambda P^{\mathrm{T}}) \mathbf{x}_t \\
&= P \Lambda^2 P^{\mathrm{T}} \mathbf{x}_t,
\end{aligned}$$

since by orthogonality $P^{\mathrm{T}} P = I$. In general we have

$$\mathbf{x}_{t+s} = P \Lambda^s P^{\mathrm{T}} \mathbf{x}_t.$$

It is obvious that for the system to converge the origin, each of the eigenvalues must be less than one in absolute value. In other words, the **spectral radius** of A, defined

as the largest of the absolute values of the eigenvalues, must be less than one. We
have shown the following result.[4]

Theorem 9.2 *Suppose that the transition function f is linear and all eigenvalues
of f have absolute values less than one. Then*

$$\lim_{t \to \infty} f^t(\mathbf{x}) = \mathbf{0}$$

for all $\mathbf{x} \in \mathbb{R}^n$.

In an economic system, $\mathbf{x}_t$ often represents the deviations of a vector of variables
from their steady-state values. Convergent to the origin therefore means that the
system converges to a steady-state equilibrium.

Example 9.6 Suppose $\mathbf{x}_{t+1} = A\mathbf{x}_t$ where

$$A = \frac{1}{4}\begin{pmatrix} 5 & 3 \\ 3 & 5 \end{pmatrix}.$$

The eigenvalues are $\lambda_1 = 2$ and $\lambda_2 = 1/2$, with corresponding eigenvectors $\mathbf{v}_1 =
(1/\sqrt{2}, 1/\sqrt{2})$ and $\mathbf{v}_2 = (-1/\sqrt{2}, 1/\sqrt{2})$. In Fig. 9.3, S_1 and S_2 are the subspaces
spanned by the vectors $\mathbf{v}_1$ and $\mathbf{v}_2$ respectively. A vector $\mathbf{x}_t$ is projected onto S_1 and
S_2. Its coordinate in S_1 is then scaled by the eigenvalue $\lambda_1 = 2$. Similarly, the

Fig. 9.3 A linear dynamical
system

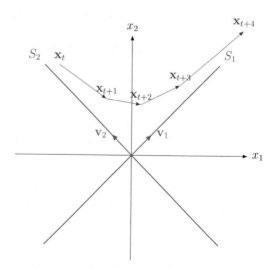

---

[4]Theorem 9.2 applies to asymmetric linear functions as well. See Devaney (2003) or Hasselblatt
and Katok (2003, Chapter 3) for details.

coordinate in S_2 is scaled by $\lambda_2 = 1/2$. In the next period $\mathbf{x}_{t+1}$ repeats the same transformation. Figure 9.3 depicts the forward orbit of such a point. The sequence is expanding by a factor of 2 in the $\mathbf{v}_1$ direction and contracting by a factor of $1/2$ in the $\mathbf{v}_2$ direction. The readers are encouraged to repeat the process starting from difference points. The resulting map will become a trajectory map of the system.

Notice that in the above example S_1 and S_2 are **invariant subspaces**, that is, once an orbit enter one of the subspaces, it cannot escape since $A\mathbf{x} = \lambda\mathbf{x}$. Formally, a set S is invariant under a transition function f if $\mathbf{x}_t \in S$ implies that $\mathbf{x}_{t+s} \in S$ for all $s \geq 1$. Moreover, S_1 is called an unstable subspace since any forward orbit in it diverges. On the other hand, S_2 is a stable subspace since any forward orbit entering it converges to the origin.

9.4.2 Nonlinear Systems

Suppose that the transition function $f : X \rightarrow X$, where X is a convex compact set in $\mathbb{R}^n$, is differentiable. By the Brouwer fixed point theorem f has at least one fixed point. Our interest is to find out orbits in the neighbourhood of each fixed point converge to or diverge from it. It turns out that the behaviour of a nonlinear system is similar to a linear system with some generalizations.

Let $\mathbf{p}$ be a fixed point of f, that is, $f(\mathbf{p}) = \mathbf{p}$. Let $Df(\mathbf{p})$ be the derivative of f evaluated at $\mathbf{p}$, which is an $n \times n$ matrix but not necessarily symmetric. Then $\mathbf{p}$ is defined as

1. a **sink** or **attracting point** if all the eigenvalues of $Df(\mathbf{p})$ are less than one in absolute value,
2. a **source** or **repelling point** if all the eigenvalues of $Df(\mathbf{p})$ are greater than one in absolute value,
3. a **saddle point** if some eigenvalues of $Df(\mathbf{p})$ are less than one and some are greater than one in absolute value.

Theorem 9.3 *Suppose f has sink at $\mathbf{p}$. Then there exists an open set U with $\mathbf{p} \in U$ in which the forward orbit of any point converges to $\mathbf{p}$.*

The largest of such open set is called the **basin of attraction**. The proof of Theorem 9.3 for the case of $\mathbb{R}^2$ can be found in Devaney (2003, p. 216). A corollary of the theorem is that if $\mathbf{p}$ is a source and f is invertible, then there exists an open set in which the backward orbit of any point converges to $\mathbf{p}$.

The case for a saddle point is more interesting. Recall the example of a linear transition function in Sect. 9.4.1 with eigenvalues $\lambda_1 = 2$ and $\lambda_2 = 1/2$. Therefore $\mathbf{p} = \mathbf{0}$ is a saddle point. There we find a stable and an unstable subspaces, spanned by the eigenvectors of λ_2 and λ_1 respectively. The equivalent invariant subsets in the nonlinear case are not linear subspaces anymore but manifolds. Manifolds are high-dimensional analogues of smooth curves (one-dimensional manifolds) or surfaces (two-dimensional manifolds). A stable manifold is often called a **saddle path**. Another example will illustrate the point.

Example 9.7 Suppose that $X \subseteq \mathbb{R}^2$ and

$$f(x, y) = \left(\frac{x}{2}, \ 2y - \frac{15}{8}x^3\right).$$

Then $f(0) = 0$ so that $\mathbf{p} = 0$ is a fixed point. The derivative is

$$Df(x, y) = \begin{bmatrix} 1/2 & 0 \\ -\frac{45}{8}x^2 & 2 \end{bmatrix},$$

and hence

$$Df(0) = \begin{bmatrix} 1/2 & 0 \\ 0 & 2 \end{bmatrix}.$$

Thus we have a saddle point at the origin. Since $f(0, t) = (0, 2t)$, the y-axis is the unstable manifold. Also,

$$f(t, t^3) = \left(\frac{t}{2}, \ 2t^3 - \frac{15}{8}t^3\right)$$

$$= \left(\frac{t}{2}, \ \left(\frac{t}{2}\right)^3\right),$$

so that $y = x^3$ is the stable manifold. Any orbit entering the saddle path $y = x^3$ will converge to the origin. Otherwise it will diverge.

In general, for an n-dimensional system, if $Df(\mathbf{p})$ has k number of eigenvalues having absolute values less than one and $n - k$ eigenvalues with absolute values greater than one, then there is a k-dimensional stable manifold that forward orbits converge to the fixed point $\mathbf{p}$. Also, the unstable manifold will be of dimension $n - k$, in which backward orbits, if f is invertible, converge to $\mathbf{p}$.

9.4.3 Phase Diagrams

An useful device for low-dimensional systems up to $\mathbb{R}^3$ is the construction of a phase diagram. It can provide useful visual insights to the overall behaviour of the system, particularly for the case of a saddle point.

In period t a point will move from $\mathbf{x}_t$ to $\mathbf{x}_{t+1}$. Therefore the vector $\Delta\mathbf{x}_{t+1} = \mathbf{x}_{t+1} - \mathbf{x}_t$ gives the direction and magnitude of the move from $\mathbf{x}_t$. Given a dynamical system $f : X \to X$, the function $d : X \to X$ where $d(\mathbf{x}) = \Delta\mathbf{x} = f(\mathbf{x}) - \mathbf{x}$ is called a **vector field**. Of course **p** is a fixed point if and only if $d(\mathbf{p}) = \mathbf{0}$. A phase diagram is a graph showing the vector field of a dynamical system.

The first step is to identify the regions where the vector field is increasing, that is, $\Delta \mathbf{x} \geq \mathbf{0}$. The complement of these regions is where $\Delta \mathbf{x} \ll \mathbf{0}$. With this information the approximate locations of the stable manifold can be estimated.

Example 9.8 Consider again the linear map

$$A = \frac{1}{4}\begin{pmatrix} 5 & 3 \\ 3 & 5 \end{pmatrix}.$$

For notational convenience let $\mathbf{x} = (x, y)^{\mathrm{T}}$. Then

$$\Delta \mathbf{x} = A\mathbf{x} - \mathbf{x}$$
$$= (A - I)\mathbf{x}$$

or

$$\begin{bmatrix} \Delta x \\ \Delta y \end{bmatrix} = \frac{1}{4}\begin{bmatrix} 1 & 3 \\ 3 & 1 \end{bmatrix}\begin{bmatrix} x \\ y \end{bmatrix}.$$

Thus $\Delta x \geq 0$ implies that $x + 3y \geq 0$ or $y \geq -x/3$. Figure 9.4a depicts the graph of $\Delta x = x + 3y = 0$, which is called a **phase line**. Points in the region above the line are increasing in the horizontal direction in their forward orbits, while points below are decreasing. On the other hand, $\Delta y \geq 0$ implies that $3x + y \geq 0$ or $y \geq -3x$. The phase line is shown in Fig. 9.4b, with points above the line going upward in their forward orbits and points below going downward. Putting the two phase lines together as in Fig. 9.4c, we can infer the directions of the forward orbits in different regions. The location of the saddle path can be approximated using the information, which lies in the stable subspace in the direction of $\mathbf{v}_2$ in Fig. 9.3.

In a linear system with a saddle point, a phase line is a hyperplane passing through the fixed point. The hyperplane divides the whole space into two half-spaces, with orbits going in opposite directions.

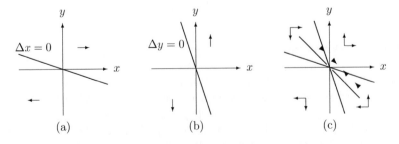

Fig. 9.4 Phase diagram of a linear system

Example 9.9 Consider the Ramsey model we analysed in Examples 9.2 and 9.3. We have $\mathbf{x} = [k \ c]^{\mathrm{T}}$. The transition equations are the capital accumulation equation (9.11) and the Euler equation (9.12) of intertemporal consumption. In the steady state, $k = k^*$ and $c = c^*$. The Euler equation reduces to $F'(k^*) = \theta + \delta$ so that

$$k^* = (F')^{-1}(\theta + \delta).$$

From the capital accumulation equation we have

$$c^* = F(k^*) - \delta k^*.$$

Using Eq. (9.11), $\Delta k \geq 0$ implies that

$$c \leq F(k) - \delta k.$$

With the usual assumptions on the production function F (the Inada condition), the phase line $\Delta k = 0$ has the shape shown in Fig. 9.5a. Capital stock above this phase line is decreasing and below increasing. From the Euler equation,

$$\Delta c \geq 0 \Leftrightarrow c_{t+1} \geq c_t$$

$$\Leftrightarrow U'(c_{t+1}) \leq U'(c_t)$$

$$\Leftrightarrow \beta(F'(k_{t+1}) + 1 - \delta) \geq 1$$

$$\Leftrightarrow F'(k_{t+1}) \geq \theta + \delta$$

$$\Leftrightarrow k_{t+1} \leq (F')^{-1}(\theta + \delta)$$

$$\Leftrightarrow k_{t+1} \leq k^*.$$

Since $k_{t+1} = F(k_t) + (1 - \delta)k_t - c_t$, the above result means that $\Delta c \geq 0$ if and only if

$$F(k) + (1 - \delta)k - c \leq k^*,$$

or

$$c \geq F(k) + (1 - \delta)k - k^*.$$

Figure 9.5b shows the phase line $\Delta c = 0$. Consumption above this phase line is increasing and below decreasing. Combining all the information in Fig. 9.5c, we see that the stable manifold or saddle path converges to the steady-state equilibrium (k^*, c^*) in the regions between the two phase lines.

In practice, with nonlinear functional forms for U and F, an explicit formulation for the stable manifold may not be possible. Many analysts choose to linearize the

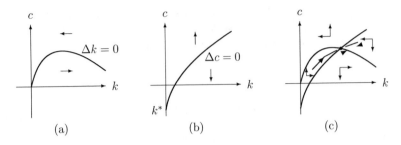

Fig. 9.5 Phase diagram of the Ramsey model

system about the fixed point (k^*, c^*). Others may solve the nonlinear system by using numerical methods.

The theory of dynamical systems has become an important tool in economic modelling. Dynamic general equilibrium models rely on a linearized system converging to a unique fixed point. This assumption can be too restrictive and richer and more interesting models can be achieved by considering nonlinear systems. For an introduction to the basic theory see Day (1994), Devaney (2003) and Hasselblatt and Katok (2003). For applications in economics see Azariadis (1993) and Day (1999).

9.4.4 Stochastic Processes: Once More with Feeling

In this section we introduce two applications of dynamical systems in stochastic processes. The first model, vector autoregression, is a useful tool in macroeconometrics. The second model, Markov chain, is an important ingredient in DSGE models. First we define the idea of stationarity. A stochastic process is said to be **covariance stationary** if it satisfies two properties:

1. The mean is independent of time, $E[\mathbf{x}_t] = \boldsymbol{\mu}$ for every t.
2. The covariance matrix between two time periods, $E[(\mathbf{x}_{t+s} - \boldsymbol{\mu})(\mathbf{x}_t - \boldsymbol{\mu})^{\mathsf{T}}]$, depends on the separation of time $s = 0, 1, 2, \ldots$, but not on t.

Vector Autoregression
Some economists suggest that the stock market is a leading indicator for the aggregate economy, but not the other way round.[5] Moreover, equity returns are by themselves random walks. Let $\mathbf{x}_t = [r_t \ \gamma_t]^{\mathsf{T}}$ be the rate of return of the stock market and growth rate of the economy respectively in period t and $\mathbf{c} = [c_1 \ c_2]^{\mathsf{T}}$ be their average rates in the long run. The relation can be represented by the model

---

[5] See, for example, Economist (2011).

$$\begin{bmatrix} r_{t+1} \\ \gamma_{t+1} \end{bmatrix} = \begin{bmatrix} c_1 \\ c_2 \end{bmatrix} + \begin{bmatrix} \phi_{11} & \phi_{12} \\ \phi_{21} & \phi_{22} \end{bmatrix} \begin{bmatrix} r_t \\ \gamma_t \end{bmatrix} + \begin{bmatrix} \epsilon_{t1} \\ \epsilon_{t2} \end{bmatrix},$$

where ϵ_{t1} and ϵ_{t2} are white noises. The above ideas can be tested by the hypotheses $\phi_{11} = \phi_{12} = 0$ and $\phi_{21} > 0$.

In general, let $\mathbf{x}_t$ be a state variable in period t and $\mathbf{c}$ be a constant vector, both in $\mathbb{R}^n$. Suppose ϵ_t be an n-dimensional vector of white noise, that is,

1. $E[\epsilon_t] = \mathbf{0} \in \mathbb{R}^n$,
2. $E[\epsilon_t \epsilon_s^\mathrm{T}] = \mathbf{0} \in \mathbb{R}^n \times \mathbb{R}^n$ for $t \neq s$,
3. $E[\epsilon_t \epsilon_t^\mathrm{T}] = \Omega$,

where Ω is an $n \times n$ symmetric positive definite covariance matrix. Then a **first-order vector autoregressive process**, VAR(1), is given by

$$\mathbf{x}_t = \mathbf{c} + \Phi \mathbf{x}_{t-1} + \epsilon_t, \tag{9.25}$$

where Φ is an $n \times n$ transition matrix. It is clear that this is a multivariate version of the AR(1) process described in Sect. 8.7.3. In view of Theorem 9.2 and Exercise 15 below, the process is covariance stationary if all the eigenvalues of Φ have absolute values less than one. That is, $\mathbf{x}_t$ has constant mean and variance in every period. The expected values of $\mathbf{x}_t$ in Eq. (9.25) is

$$E[\mathbf{x}_t] = \mathbf{c} + \Phi E[\mathbf{x}_{t-1}] + E[\epsilon_t],$$

which gives

$$\mu = E[\mathbf{x}_t] = E[\mathbf{x}_{t-1}] = (I_n - \Phi)^{-1}\mathbf{c}.$$

Finding the covariance matrix Σ of $\mathbf{x}_t$ is more involving but it can be shown that Σ satisfies the following discrete Lyapunov matrix equation:

$$\Sigma = \Phi \Sigma \Phi^\mathrm{T} + \Omega. \tag{9.26}$$

See Stock and Watson (2001) for an introduction in applying VAR to macroeconomics. Hamilton (1994) and Lütkepohl (2005) provide extensive coverage on analysis and estimation of VAR models.

Markov Chains

Consider a typical worker in an economy. Suppose that if the worker is employed in this period, then the probability of staying employed in the next period is p_{11}. The probability of unemployment in the next period is therefore $p_{12} = 1 - p_{11}$. On the other hand, if the worker is unemployed in this period and the chance of finding a job in the next period is p_{21}, the chance of remaining unemployed is then

$p_{22} = 1 - p_{21}$. We can represent the status $\mathbf{x}_t$ of the worker with the standard unit vectors in $\mathbb{R}^2$. That is, $\mathbf{x}_t = \mathbf{e}_1 = [1\ 0]^T$ means that the worker is employed in period t, and $\mathbf{x}_t = \mathbf{e}_2 = [0\ 1]^T$ unemployed. The conditional probabilities defined above are therefore

$$p_{ij} = \text{Prob}(\mathbf{x}_{t+1} = \mathbf{e}_j | \mathbf{x}_t = \mathbf{e}_i), \quad i, j = 1, 2.$$

We can summarize the situation by a transition matrix

$$P = \begin{bmatrix} p_{11} & p_{12} \\ p_{21} & p_{22} \end{bmatrix}. \tag{9.27}$$

Now imagine we draw a random person from the work force in period t and let $\boldsymbol{\pi}_t = [\pi_{t1}\ \pi_{t2}]^T$ be the probabilities that she is employed or unemployed respectively. Here $\pi_{t2} = 1 - \pi_{t1}$ is the unemployment rate of the economy in period t. The probability distribution of this person's employment status in the next period is then

$$\begin{bmatrix} \pi_{(t+1)1} \\ \pi_{(t+1)2} \end{bmatrix} = \begin{bmatrix} p_{11} & p_{21} \\ p_{12} & p_{22} \end{bmatrix} \begin{bmatrix} \pi_{t1} \\ \pi_{t2} \end{bmatrix},$$

or

$$\boldsymbol{\pi}_{t+1} = P^T \boldsymbol{\pi}_t.$$

Suppose that the average income of an employed worker is $\bar{y}_1$ and that of an unemployed person is $\bar{y}_2$. Write $\bar{\mathbf{y}} = [\bar{y}_1\ \bar{y}_2]^T$. Then $y_t = \bar{\mathbf{y}}^T \mathbf{x}_t$ is a random variable of income in period t. To summarize, $\{y_t\}$ is a stochastic process of labour income with probability distribution $\boldsymbol{\pi}_t$ in period t.

In the general set up, the state of the world consists of n possible outcomes, denoted by the state vectors $\mathbf{x}_t \in \{\mathbf{e}_1, \dots, \mathbf{e}_n\}$. The conditional probability that state j will happen in period $t + 1$ given that state i occurs in period t is

$$p_{ij} = \text{Prob}(\mathbf{x}_{t+1} = \mathbf{e}_j | \mathbf{x}_t = \mathbf{e}_i), \quad i, j = 1, \dots, n.$$

The $n \times n$ probabilities are conveniently written as the transition or **stochastic matrix**

$$P = \begin{bmatrix} p_{11} & p_{12} & \cdots & p_{1n} \\ \vdots & \vdots & & \vdots \\ p_{n1} & p_{n2} & \cdots & p_{nn} \end{bmatrix}, \tag{9.28}$$

with the properties that all elements are non-negative and all the rows sum to one:

$$\sum_{j=1}^{n} p_{ij} = 1, \quad i = 1, \dots, n.$$

The probability distribution of the n states in each period t is given by an $n \times 1$ vector $\boldsymbol{\pi}_t = [\pi_{t1} \cdots \pi_{tn}]^{\mathrm{T}}$, with $\sum_{i=1}^{n} \pi_{ti} = 1$. The transition equation for the probability distribution is therefore[6]

$$\boldsymbol{\pi}_{t+1} = P^{\mathrm{T}} \boldsymbol{\pi}_t.$$

This is an autonomous linear dynamical system on the simplex

$$S = \left\{ \boldsymbol{\pi} \in \mathbb{R}^n : \pi_i \geq 0, \sum_{i=1}^{n} \pi_i = 1 \right\}. \tag{9.29}$$

It can be shown (see Exercise 23 below) that P^{T} has at least one eigenvalue equal to one. That is, there exists an eigenvector $\boldsymbol{\pi} \in S$ such that

$$\boldsymbol{\pi} = P^{\mathrm{T}} \boldsymbol{\pi}, \tag{9.30}$$

which implies that $\boldsymbol{\pi}$ is a fixed point, or in this case called a stationary probability distribution of the Markov chain. A Markov chain is called **regular** if some power of the transition matrix has only positive elements. That is, there is an integer k such that P^k is a positive matrix. It turns out that for a regular Markov chain, all initial probability distributions converge to one unique fixed point. The process is said to be **asymptotically stationary**, with

$$\lim_{t \to \infty} P^t = \Pi,$$

where all the n rows of Π are identical and equal to $\boldsymbol{\pi}$.

Now we derive the stationary probability distribution $\boldsymbol{\pi}$ of a regular Markov chain. Since any probability distribution must sum to one, $\sum_{1=1}^{n} \pi_i = 1$, or, if we define $\mathbf{1}$ to be an n-vector of ones,

$$\mathbf{1}^{\mathrm{T}} \boldsymbol{\pi} = 1. \tag{9.31}$$

Also, Eq. (9.30) implies that

$$(I_n - P^{\mathrm{T}})\boldsymbol{\pi} = \mathbf{0}. \tag{9.32}$$

Combining Eqs. (9.31) and (9.32), we have

$$A\boldsymbol{\pi} = \mathbf{e}_1, \tag{9.33}$$

---

[6]Some authors prefer writing $\boldsymbol{\pi}_t$ as a row vector so that the transition equation is $\boldsymbol{\pi}_{t+1} = \boldsymbol{\pi}_t P$. Others define the transition matrix P as the transpose of what we specified in Eq. (9.28), consequently the transition equation is $\boldsymbol{\pi}_{t+1} = P\boldsymbol{\pi}_t$.

where $\mathbf{e}_1$ is the $(n+1)$-dimensional standard unit vector, and

$$A = \begin{bmatrix} \mathbf{1}^\mathrm{T} \\ I_n - P^\mathrm{T} \end{bmatrix}$$

is an $(n+1) \times n$ matrix. Premultiplying both sides of Eq. (9.33) by $(A^\mathrm{T}A)^{-1}A^\mathrm{T}$, we get

$$\boldsymbol{\pi} = (A^\mathrm{T}A)^{-1}A^\mathrm{T}\mathbf{e}_1.$$

That is, $\boldsymbol{\pi}$ is the first column of the matrix $(A^\mathrm{T}A)^{-1}A^\mathrm{T}$.

Let $\bar{\mathbf{y}} \in \mathbb{R}^n$ be given. Then the inner product $y_t = \bar{\mathbf{y}}^\mathrm{T}\mathbf{x}_t$ defines a real random variable in period t with probability distribution π_t so that $\{y_t\}$ is a stochastic process. Given a Markov chain with initial probability distribution $\boldsymbol{\pi}_0$, the distribution after t periods is $\boldsymbol{\pi}_t = (P^\mathrm{T})^t \boldsymbol{\pi}_0$. The expected value of y_t is therefore

$$E[y_t] = \boldsymbol{\pi}_t^\mathrm{T}\bar{\mathbf{y}} = \boldsymbol{\pi}_0^\mathrm{T}P^t\bar{\mathbf{y}}.$$

See Grinstead and Snell (1997) and Hamilton (1994) for introductions to Markov chains; Levin, Peres, and Wilmer (2009) provides a more advanced treatment.

9.5   Exercises

1. Draw the graphs showing the dynamics of Eq. (9.1) for the cases of $-1 < a < 0$ and $a < -1$.
2. Prove statement (9.5).
3. Give an economic interpretation to the limiting cases $\lambda = 0$ and $\lambda = 1$ in Eq. (9.7).
4. Prove that $\mathrm{Fix}(f)$ is a closed set if f is continuous.
5. Consider an increasing and continuous function $f : X \to X$ where $X \subseteq \mathbb{R}$ is a compact set. From Exercise 4 above the set $\{x \in X : f(x) \neq x\}$ is an open set and therefore a collection of open intervals. Take an interval $I = (a, b)$ where $f(a) = a$, $f(b) = b$ and $f(x) \neq x$ for $a < x < b$. Show that every forward orbit in I converges to b if $f(x) > x$ and converges to a if $f(x) < x$.
6. A consumer's savings problem is

$$\max_{c_t} E_0 \sum_{t=0}^{\infty} \beta^t U(c_t),$$

subject to the budget constraint

$$a_{t+1} = (1 + r_t)(a_t + x_t - c_t),$$

where c_t is consumption, a_t is asset holding, x_t is an exogenous income, r_t is the rate of return on asset and $\beta = 1/(1+\theta)$.

(a) Set up the Bellman equation and show that the Euler equation is

$$\beta E_t (1 + r_{t+1}) \frac{U'(c_{t+1})}{U'(c_t)} = 1.$$

(b) Suppose that utility is represented by the quadratic function $U(c_t) = -(c_t - \gamma)^2$ and $r_t = \theta$ for all periods. Show that consumption is a martingale, that is, $E_t c_{t+1} = c_t$.

7. Consider the lump-sum taxation model where the central planer maximizes household utility

$$\sum_{s=0}^{\infty} \beta^s \left(c_{t+s}^{\alpha} g_{t+s}^{1-\alpha} \right), \quad 0 < \alpha < 1,$$

subject to the resource constraint

$$F(k_t) = c_t + k_{t+1} - (1 - \delta)k_t + g_t$$

and the government budget constraint $g_t = T_t$.

(a) Set up the Bellman equation.
(b) Derive the necessary conditions from the Bellman equation.
(c) Find the Euler equation, the marginal rate of substitution of household consumption to government services, and the steady-state condition.

8. Consider the following cake-eating problem:

$$\max_{c_t} \sum_{t=1}^{\infty} \beta^t U(c_t, c_{t-1}),$$

subject to $w_{t+1} = w_t - c_t$ with the initial size of the cake w_0 given. In each period the consumer eat part of the cake (c_t) but must save the rest.

(a) Set up the Bellman equation.
(b) Derive the Euler equation.

9. Verify each step in Example 9.4.

10. (Ljungqvist and Sargent 2004, p. 110) Consider the linear quadratic model in Sect. 9.3.3 with $\beta = 1$. Suppose we know that the value function is also a quadratic form, $v(x) = -x^T P x$, where P is a positive semidefinite symmetric matrix.

(a) Set up the Bellman equation using this information.
(b) Show that

$$P = R + A^T P A - A^T P B \left(Q + B^T P B \right)^{-1} B^T P A. \tag{9.34}$$

Equation (9.34) is called a Riccati equation in P and requires numerical analysis to solve.

11. In a linear dynamical system in $\mathbb{R}^2$ where $|\lambda_1| > 1$ and $|\lambda_2| < 1$, what is the necessary condition for the system to converge to the origin?

12. Solve the second-order difference equation (9.8).

13. Describe the dynamics of the following linear maps with matrix representations

$$\text{(a)} \begin{pmatrix} 2 & 1 \\ 1 & 1 \end{pmatrix}, \quad \text{(b)} \begin{pmatrix} -\frac{1}{2} & 0 \\ 0 & 2 \end{pmatrix},$$

$$\text{(c)} \ \beta \begin{pmatrix} \cos\alpha & -\sin\alpha \\ \sin\alpha & \cos\alpha \end{pmatrix},$$

with $0 \le \alpha < 2\pi$ and $\beta > 0$.

14. Suppose that all the eigenvalues of a square matrix A have absolute values less than one.
    (a) Show that the matrix $I - A$ is invertible. (Hint: Let $(I - A)\mathbf{x} = \mathbf{0}$, where $\mathbf{x}$ is a column vector. Show that $\mathbf{x} = \mathbf{0}$.)
    (b) Prove that $(I - A)^{-1} = I + A + A^2 + \cdots$.
    Note: This is a multivariate version of Eq. (3.3).

15. Suggest a solution for the affine dynamical system

$$\mathbf{x}_{t+1} = A\mathbf{x}_t + \mathbf{b},$$

where $\mathbf{b} \in \mathbb{R}^n$. What is the necessary condition for the system to converge to a fixed point? (Hint: This is a multivariate version of the first-order difference equation discussed in Sect. 9.1.1.)

16. Derive the equations for the phase lines for the dynamical system in the example of Sect. 9.4.2:

$$f(x, y) = \left(\frac{x}{2}, \ 2y - \frac{15}{8}x^3 \right).$$

Draw the phase diagram and locate the stable and unstable manifolds.

17. Consider the one-dimensional Solow-Swan model

$$\Delta k_{t+1} = sF(k_t) - (\delta + n)k_t,$$

where k_t is the capital stock in period t, and s, δ and n are the exogenous rates of saving, depreciation and population growth, respectively. The production function F satisfies the Inada conditions.
    (a) Find the set of fixed points.
    (b) What are the necessary conditions for the system to converge?
    (c) Draw a diagram showing the dynamics of k.
    (d) Find the basins of attraction.

18. Growth in the Solow-Swan model is forced by the strong assumptions on the production function. Here is another model that is more realistic at low income

levels. Let x_t be the income per person of an economy at period t. Suppose that $x_{t+1} = f(x_t)$ where

$$f(x) = \begin{cases} (b-a)^{1-\alpha}(x-a)^\alpha + a, & \text{if } a \le x \le b, \\ (x-b)^\beta + b, & \text{if } x > b. \end{cases}$$

In the above, $0 < a < b, \alpha > 1$ and $0 < \beta < 1$.
(a) Find Fix(f).
(b) Find the basin of attraction for each fixed point.
The region $a \le x \le b$ is called a poverty trap. See Fig. 9.6 and Banerjee and Duflo (2011, p. 12). The book provides justifications for the shape of the growth curve.

19. Consider a stationary VAR(1) process as in Eq. (9.25). Let $\bar{\mathbf{x}}_t = \mathbf{x}_t - \boldsymbol{\mu}$, the derivation of $\mathbf{x}_t$ from the mean.
    (a) Show that the process can be represented by

$$\bar{\mathbf{x}}_t = \Phi\bar{\mathbf{x}}_{t-1} + \boldsymbol{\epsilon}_t.$$

    (b) As in the theory of ordinary least square, the error term $\boldsymbol{\epsilon}_t$ is assumed to be uncorrelated with $\bar{\mathbf{x}}_{t-1}$, that is, $E[\bar{\mathbf{x}}_{t-1}\boldsymbol{\epsilon}_t^T] = \mathbf{0} \in \mathbb{R}^n \times \mathbb{R}^n$. Show that the covariance matrix Σ of $\mathbf{x}_t$ satisfies Eq. (9.26).

20. Consider the following statement:

    The Bureau of Labour Statistics estimates that workers recently laid off have a 30% chance of finding work in a given month. For workers off the job for more than six months, that chance is no better than 10%.[7]

    Can the employment dynamics be modelled by a Markov chain? Why or why not? If yes, what will the transition matrix look like?

Fig. 9.6 The poverty trap

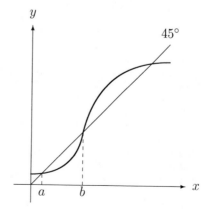

---

[7]"Six Years into a Lost Decade", *The Economist*, August 6, 2011.

21. Let $(P, \mathbf{x})$ be an n-state Markov chain. Show that

$$E[\mathbf{x}_{t+1}|\mathbf{x}_t] = P^{\mathrm{T}}\mathbf{x}_t.$$

22. Show that the eigenvalues of the two-state Markov chain defined in Eq. (9.27) are $\lambda_1 = 1$ and $\lambda_2 = p_{11} + p_{22} - 1$.

23. Let P be an $n \times n$ matrix (not necessary symmetric) which has nonnegative elements and satisfies $\sum_{j=1}^{n} p_{ij} = 1$ for $i = 1, \ldots, n$. Show that P^{T} has at least one unit eigenvalue.
    Hint: Consider the simplex defined in Eq. (9.29). Show that the linear function f represented by P^{T} is a linear operator on S, that is, for every $\boldsymbol{\pi} \in S$, $f(\boldsymbol{\pi})$ is also in S. Then show that $f : S \to S$ satisfies the hypothesis of the Brouwer Fixed Point Theorem.

References

Adda, J., & Cooper, R. (2003). *Dynamic economics*. Cambridge: The MIT Press.

Azariadis, C. (1993). *Intertemporal macroeconomics*. Cambridge: Blackwell Publishers.

Banerjee, A., & Duflo, E. (2011). *Poor economics: A radical rethinking of the way to fight global poverty*. New York: PublicAffairs.

Boyd, S. (2008). *Introduction to linear dynamical systems*. Lecture Notes for EE263, Stanford University. (Lectures available on iTune U)

Carter, M. (2001). *Foundations of mathematical economics*. Cambridge: The MIT Press.

Conrad, J. M. (2010). *Resource economics*, Second edition. Cambridge: Cambridge University Press.

Grinstead, C. M., & Snell, J. L. (1997). *Introduction to probability*, Second Revised Edition. Providence: American Mathematical Society. (Downloadable from the Chance web site).

Day, R. H. (1994). *Complex economic dynamics*, Volume I. Cambridge: The MIT Press.

Day, R. H. (1999). *Complex economic dynamics*, Volume II. Cambridge: The MIT Press.

Devaney, R. L. (2003). *An introduction to chaotic dynamical systems*, Second Edition. Cambridge: Westview Press.

Economist. (2011). The Missing Link, May 19 issue.

Goldberg, S. (1986). *Introduction to difference equations*. New York: Dover Publications.

Hamilton, J. D. (1994). *Time series analysis*. Princeton: Princeton University Press.

Hasselblatt, B., & Katok, A. (2003). *A first course in dynamics: with a panorama of recent developments*. Cambridge: Cambridge University Press.

Levin, D. A., Peres, Y., & Wilmer, E. L. (2009). *Markov chains and mixing times*. Providence: American Mathematical Society. (Chapters 1 and 2 downloadable from the AMS web site).

Ljungqvist, L., & Sargent, T. J. (2004). *Recursive macroeconomic theory*, Second Edition. Cambridge: The MIT Press.

Lütkepohl, H. (2005). *New introduction to multiple time series analysis*. Berlin: Springer-Verlag.

Stock, J. H., & Watson, M. W. (2001). Vector autoregressions. *Journal of Economic Perspectives*, *15*(4), 101–115.

Stokey, N. L., & Lucas, R. E. (1989). *Recursive methods in economic dynamics*. Cambridge: Harvard University Press.

Weitzman, M. L. (2003). *Income, wealth, and the maximum principle*. Cambridge: Harvard University Press.

Wickens, M. (2011). *Macroeconomic theory: A dynamic general equilibrium approach*, Second Edition. Princeton: Princeton University Press.

Bibliography

Adda, J., & Cooper, R. (2003). *Dynamic economics*. Cambridge: The MIT Press.

Artin, E. (1964). *The gamma function*. New York: Holt, Rinehart and Winston.

Azariadis, C. (1993). *Intertemporal macroeconomics*. Cambridge: Blackwell Publishers.

Banerjee, A., & Duflo, E. (2011). *Poor economics: A radical rethinking of the way to fight global poverty*. New York: PublicAffairs.

Beck, M., & Geoghegan, R. (2010). *The art of proof*. New York: Springer Science+Business Media.

Boyd, S. (2008). *Introduction to linear dynamical systems*. Lecture Notes for EE263, Stanford University. (Lectures available on iTune U)

Box, G. E. P., Jenkins, G. M., & Reinsel, G. C. (1994). *Time series analysis: forecasting and control*, Third edition. Englewood Cliffs: Prentice-Hall, Inc.

Brosowski, B., & Deutsch, F. (1981). An elementary proof of the Stone-Weierstrass theorem. *Proceedings of the American Mathematical Society, 81*(1), 89–92.

Carter, M. (2001). *Foundations of mathematical economics*. Cambridge: The MIT Press.

Clausen, A. (2006). *Log-linear approximations*. Unpublished class note, University of Pennsylvania.

Conrad, J. M. (2010). *Resource economics*, Second edition. Cambridge: Cambridge University Press.

Day, R. H. (1994). *Complex economic dynamics*, Volume I. Cambridge: The MIT Press.

Day, R. H. (1999). *Complex economic dynamics*, Volume II. Cambridge: The MIT Press.

Debreu, G. (1952). Definite and semidefinite quadratic forms. *Econometrica, 2*(20), 295–300.

Devaney, R. L. (2003). *An introduction to chaotic dynamical systems*, Second Edition. Cambridge: Westview Press.

Devlin, K. (1993). *The joy of sets*, Second Edition. New York: Springer Science+Business Media.

Devlin, K. (2002). Kurt Gödel—separating truth from proof in mathematics. *Science*, Volume 298, December 6 issue, 1899–1900.

Diewert, W. E. (1999). *Unconstrained optimization*. Unpublished Lecture Notes, University of British Columbia.

Economist. (1999). Getting the Goat. February 18 issue.

Economist. (2008). Easy as 1, 2, 3. December 30 issue.

Economist. (2011). The Missing Link. May 19 issue.

Ellis, G. (2012). On the philosophy of cosmology. Talk at Granada Meeting, 2011. Available at <http://www.mth.uct.ac.za/~ellis/philcosm_18_04_2012.pdf>.

Gerstein, L. J. (2012). *Introduction to mathematical structures and proofs*, Second Edition. New York: Springer Science+Business Media.

Goldberg, S. (1986). *Introduction to difference equations*. New York: Dover Publications.

Grinstead, C. M., & Snell, J. L. (1997). *Introduction to probability*, Second Revised Edition. Providence: American Mathematical Society. (Downloadable from the Chance web site).

Hamilton, J. D. (1994). *Time series analysis*. Princeton: Princeton University Press.

© Springer Nature Switzerland AG 2019

K. Yu, *Mathematical Economics*, Springer Texts in Business and Economics,

https://doi.org/10.1007/978-3-030-27289-0

Hasselblatt, B., & Katok, A. (2003). *A first course in dynamcis: with a panorama of recent developments*. Cambridge: Cambridge University Press.

Hoffman, K., & Kunze, R. (1971). *Linear algebra*, Second Edition. Englewood Cliffs: Prentice-Hall.

Hogg, R. V., & Craig, A. T. (1995). *Introduction to mathematical statistics*, Fifth Edition. Englewood: Prentice-Hall, Inc.

Holt, J. (2008). Numbers guy. *The New Yorker*, March 3 issue.

Jehle, G. A., & Reny, P. J. (2011). *Advanced microeconomic theory*, Third edition. Harlow: Pearson Education Limited.

Johnston, J. (1991). *Econometric methods*, Third Edition. New York: McGraw-Hill.

Lancaster, K. (1968). *Mathematical economics*. New York: The Macmillan Company.

Lay, S. R. (2000). *Analysis with an introduction to proof*, Third edition. Upper Saddle River: Prentice Hall.

Levin, D. A., Peres, Y., & Wilmer, E. L. (2009). *Markov chains and mixing times*. Providence: American Mathematical Society. (Chapters 1 and 2 downloadable from the AMS web site).

Ljungqvist, L., & Sargent, T. J. (2004). *Recursive macroeconomic theory*, Second Edition. Cambridge: The MIT Press.

Lütkepohl, H. (2005). *New introduction to multiple time series analysis*. Berlin: Springer-Verlag.

Marsden, J. E., & Tromba, A. J. (1988). *Vector calculus*, Third edition. W.H. Freeman and Company.

Matso, J. (2007). Strange but true: infinity comes in different sizes. *Scientific American*, July 19 issue.

McLennan, A. (2014). *Advanced fixed point theory for economics*. Available at <http://cupid.economics.uq.edu.au/mclennan/Advanced/advanced_fp.pdf>.

Overbye, D. (2014). In the end, it all adds up to $-1/12$. *The New York Times*, February 3 issue.

Paulos, J. A. (2011). The mathematics of changing your mind. *The New York Times*, August 5 issue.

Poston, T., & Stewart, I. (1978). *Catastrophe theory and its applications*. London: Pitman.

Rao, C. R. (1973). *Linear statistical inference and its applications*, Second Edition. New York: John Wiley & Sons.

Rockafellar, R. T. (1970). *Convex analysis*. Princeton: Princeton University Press.

Rosenlicht, M. (1968). *Introduction to analysis*. Scott, Foresman and Co. (1986 Dover edition).

Ross, J. F. (2004). Pascal's legacy. *EMBO Reports*, Vol. 5, Special issue, S7–S10.

Royden, H. L., & Fitzpatrick, P. M. (2010). *Real analysis*, Boston: Prentice Hall.

Rudin, W. (1976). *Principles of mathematical analysis*, Third edition. New York: McGraw-Hill.

Starr, R. M. (2008). Shapley-Folkman theorem. *The new Palgrave dictionary of economics*, Second Edition. Houndmills: Palgrave Macmillan.

Stock, J. H., & Watson, M. W. (2001). Vector autoregressions. *Journal of Economic Perspectives, 15*(4), 101–115.

Stokey, N. L., & Lucas, R. E. (1989). *Recursive methods in economic dynamics*. Cambridge: Harvard University Press.

Strogatz, S. (2010). Division and its discontents. *The New York Times*, February 21 issue.

Sundaram, R. K. (1996). *A first course in optimization theory*. Cambridge: Cambridge University Press.

Taylor, H. M., & Karlin, S. (1998). *An introduction to stochastic modeling*, Third Edition. San Diego: Academic Press.

Timoshenko, S. (1940). *Strength of materials*. New York: D. Van Nostrand Company.

Tversky, A., & Kahneman, D. (1974). Judgment under uncertainty: heuristics and biases. *Science, 185*(4157), 1124–1131.

Weitzman, M. L. (2003). *Income, wealth, and the maximum principle*. Cambridge: Harvard University Press.

Wickens, M. (2011). *Macroeconomic theory: A dynamic general equilibrium approach*, Second Edition. Princeton: Princeton University Press.

Wiggins, C. (2006). Bayes's theorem. *Scientific American*, December 4 issue.

Index

© Springer Nature Switzerland AG 2019
K. Yu, *Mathematical Economics*, Springer Texts in Business and Economics,
https://doi.org/10.1007/978-3-030-27289-0